RANDOLPH WEMYSS
MEMORIAL HOSPITAL

Stroke

THE FACTS

BY

F. CLIFFORD ROSE
Physician in Charge

AND

RUDY CAPILDEO
Senior Registrar

Stroke Research Unit
Department of Neurology
Charing Cross Hospital
London

with a foreword by
SIR PETER MEDAWAR, FRS

OXFORD
OXFORD UNIVERSITY PRESS
NEW YORK TORONTO

Oxford University Press, Walton Street, Oxford OX2 6DP

*London Glasgow New York Toronto
Delhi Bombay Calcutta Madras Karachi
Kuala Lumpur Singapore Hong Kong Tokyo
Nairobi Dar es Salaam Cape Town
Melbourne Auckland
and associate companies in
Beirut Berlin Ibadan Mexico City*

© F. Clifford Rose and Rudy Capildeo 1981

All rights reserved. No part of this publication may be reproduced, stored in a retrieval system, or transmitted, in any form or by any means, electronic, mechanical, photocopying, recording, or otherwise, without the prior permission of Oxford University Press

This book is sold subject to the condition that it shall not, by way of trade or otherwise, be lent, re-sold, hired or otherwise circulated without the publisher's prior consent in any form of binding or cover other than that in which it is published and without a similar condition including this condition being imposed on the subsequent purchaser

British Library Cataloguing in Publication Data
Rose, F. Clifford
    Stroke. – (Oxford medical publications)
    1. Cerebrovascular disease
    I. Title   II. Capildeo, Rudy
    616.8'1   RC388.5   80-2611704
    ISBN 0-19-261170-4

*First published 1981
Reprinted 1982*

Typeset by Hope Services Ltd., Abingdon
Printed in Great Britain by
Richard Clay (The Chaucer Press) Ltd,
Bungay, Suffolk

Foreword

by

SIR PETER MEDAWAR, FRS

Some ten years ago I was the victim of a cerebral haemorrhage affecting the right side of the brain (see page 37 of this book) causing a paralysis of the left side of the body—left hemiplegia or hemiparesis (see Chapter 1). Being a writer as well as a scientist I resolved to write a book called *Stroke*, recounting my personal experiences. My ward sister did not approve of my ostensible reason for writing it—to secure specially attentive treatment from any nurse who hoped to cut a good figure in my pages. One day she came into the ward with an air of triumph, and carrying a book called *Stroke*: 'You see,' she said, 'it's been written already, so you can't'. 'Why then,' I replied, 'my book shall be called *Son of Stroke*'.

It would not have been anything like as good as this one which is a very professional job and clearly the fruit of close collaboration between neurologists and a multidisciplinary team.

A stroke is a calamity that affects not only its proximate victim (the person who actually has it) but also his relatives, friends, and even work-mates, all of whom will be disadvantaged to a greater or lesser degree by the calamity. It is not only the patient, therefore, who will derive benefit from reading this book but also his or her family and indeed anyone who wants to help the patient to return to as near an approximation as possible to a normal life. The patient however is the most important actor in the drama—as it so often turns out to be.

From this book stroke patients will learn much that will help them understand their condition, and in the course of living they will find out quite a lot the physicians may not have thought it worth while to tell them. A stroke, for example, is often accompanied by a lowered threshhold for

Foreword

emotional disturbances: the patient may be unreasonably tearful, sometimes for no apparent reason, which is sometimes interpreted as a sign of deep depression. In reality it is nothing of the kind: it may be provoked by anything that in a normal person might cause a tear to start into the eye—an act of heroism, for example, or indeed any moment of glory. The victim must thrust away any medication intended to reinstate a more reasonable frame of mind; stroke victims are in any case very often especially susceptible to the action of psychoactive drugs—drugs that alter one's mood or state of mind such as tranquillizers or sleeping tablets.

A famous American brain surgeon, a writer himself, classified the victims of strokes and the physically handicapped generally as the citizens of a 'Fourth World'. Because much the same advice applies to them all one may include in the 'Fourth World' all who are handicapped or disadvantaged by strokes of ill fortune such as the loss of a beloved wife, husband, or child; even the loss of a job may reasonably qualify. All such people must start afresh and the rubric which embodies the best advice that can be given to any is: *Adapt or perish*. A new life must be begun and the patient's address to it will affect the whole of his life. It should perhaps combine the willingness to learn and sanguine hopefulness of someone learning to drive a car.

A book that is honest and straightforward such as this is may depress a stroke victim who, as with the victims of other neurological afflictions of more stealthy onset such as multiple sclerosis, look always for encouragement and messages of hope. Fortunately one such message may now be given. The functional plasticity of the brain—a subject upon which surprising and encouraging discoveries continue to be made—is such that stroke victims who are determined to improve may do much better than might have been predicted from too literal a reading of neurological textbooks. Indeed, to do better than was expected of one is a source of pride that does something to offset the many causes of disappointment and despondency.

Preface

Strokes are the third-commonest cause of death in the Western world, after heart disease and cancer. They can occur at any age although the risk greatly increases as we get older. This fact is of universal importance since life expectancy is increasing in all countries and the chances of exceeding the biblical 'three score years and ten' are becoming much greater, even in 'underdeveloped' countries.

In addition to being a major cause of death, stroke is the single most common cause of disability, the care for which requires enormous financial resources. It is the extent of the resources available that finally determines what can be done. To the individual patient, it means loss of earning power with a greater burden on the family who will look after the disabled person at home. To those without a caring family, there may be no alternative to long-term institutional care, and the facilities for this may be limited.

This book is written for the lay person, particularly the patient, his family, and friends, in the hope that an improved understanding of the nature of the stroke illness will be of help during the period of recovery. This is often slow, when pessimism and depression are caused by uncertainty about the future rather than because of the disability caused.

Prevention of stroke is clearly of interest to us all, on a personal as well as a national level, and with the considerable research advances indicated in this book, there is hope and optimism for the future.

F. Clifford Rose
Rudy Capildeo

Contents

1. Introduction — 1
2. What is a stroke? — 5
3. How does a stroke occur? — 10
4. Risk factors — 25
5. The different types of stroke — 33
6. Investigations — 50
7. Who looks after the patient in hospital? — 59
8. Admission to hospital — 68
9. The first few days — 73
10. After the first week — 87
11. After the first month — 98
12. Surgical treatment of stroke — 103
13. The patient at home — 107
14. Assessment of the treatment of stroke — 116
15. Outcome after stroke — 124
16. The future — 132
    Glossary — 135
    Index — 139

1

Introduction

The word 'stroke' has many connotations, but when used in its medical sense, the impression conveyed is remarkably stereotyped and suggests an elderly person, whose attack occurs 'out of the blue' and is inevitably accompanied by severe and life-threatening paralysis of one half of the body.

The term derives from the ancient belief that it occurred as a stroke 'from above', usually as a punishment. This may partly account for the resigned way in which many patients and their relatives react to the sudden occurrence of any serious illness with stroke; the usual attitude is 'let's wait and see how things go over the next few days'.

The patient's story

Case 1. A 70-year-old widower, living alone after his retirement, wrote:

I had gone to bed feeling all right, got up at 6 a.m. and went to the lavatory. I felt a bit strange but went back to bed. At 7 a.m. I got up and went downstairs to make a pot of tea, as I always do. While the kettle was boiling, I took the cover off the budgie's cage. I tried to speak to him and found that my words wouldn't come out right. I made the tea but when I went to carry my cup into the living room it almost slipped from my hand. I sat in the chair for an hour and I wondered if I had suffered a slight stroke.

I managed to get dressed and walked down to the post office to get my pension. I handed my book over the counter but was frightened to say anything. The lady, who sees me every week, said that I didn't look well. I tried to answer but the words seemed to come out all wrong—I knew what I wanted to say but I couldn't find the right words. When I tried to sign the postal order, I couldn't write properly. They made me sit down for a little while and one of the staff took me to her family doctor's house. He was on a call but his wife looked after me, called a taxi and I was taken home. The doctor called on me

Introduction

as soon as he had received the message at home and came and saw me. I knew that I had had a stroke and that nothing could be done.

This patient was fortunate in that his disability was slight and he made a full recovery within three weeks of his stroke, since a more serious one would have made it impossible for him to have continued to live alone at home.

There are a number of accounts of well-known people who have suffered strokes. Samuel Johnson, the famous wit and lexicographer (an occupation which he defined as a 'harmless drudge'), described the abject fear that was caused by an episode when he suddenly lost his speech but, fortunately, from which he quickly recovered.

Strokes in world leaders have probably affected history, for example in the cases of Lenin and Franklin D. Roosevelt.

In the latter part of his life Sir Winston Churchill suffered from a number of minor strokes which affected his speech and also his hand, episodes which were fully described by his physician.

Of particular interest are those accounts from doctors who have had a stroke.

Case 2. Professor Smithells, from the University of Otago, New Zealand wrote in the *New Zealand Medical Journal* of June 1978 after his third stroke: 'I do not know who coined the word "stroke" but it is such a one-fell-swoop experience that I think stroke is a singularly appropriate term for the sudden cutting-off of competence.' Professor Smithells was formerly Director of the University's School of Physical Education. He found 'the worst blow is the loss of willpower'. He described dressing difficulties, eating problems, and the feelings of shabbiness and messiness associated with these activities. After a sleep, he felt slow to pick up and his walking seemed to be even more incoordinated.

Case 3. Professor A. Brodal, from the University of Oslo, Norway, suffered a stroke at the age of 62 years, whilst on a lecture-trip abroad. He described his illness in an analytical way, as one might expect from a Professor of Anatomy, in a paper published in the journal *Brain* in 1973. His illness 'started suddenly when I woke up and turned in bed on the morning of 12th April 1972. In the course of a few minutes an

Introduction

initial heavy, but uncharacteristic, dizziness was followed by dysarthria (slurred speech), double vision and a marked paresis (weakness) of the left arm and leg'.

Apart from Samuel Johnson's account, these histories do not indicate the real fear which is felt at the onset by stroke patients, who readily admit it on direct questioning but rarely volunteer it. This same fear re-emerges on recovery: 'Doctor, will it happen again?'

The relative's story

If the patient is not admitted to hospital, then the full burden of nursing care is borne almost entirely by those at home whose duties may encompass the full 24 hours, a day-to-day strain that can be emotionally and physically exhausting. Care at home is discussed more fully in Chapter 13, but the following account shows how much family life can be disrupted with concern for the stroke victim.

Case 4. My mother lived on her own, with her dog, in a large house. She had cut herself off from her neighbours and friends in recent years. Her two sisters and one brother lived a long way away and she rarely saw them. At the time of her stroke, she was 64 years old and it affected her speech, right hand and to some extent her right leg. She did not go to hospital but remained at home and my wife and I moved in to look after her. After 9 months she recovered completely.

Eighteen months later my mother had a second stroke which affected her left arm and left leg. We found her lying on the floor and the family doctor was called. We moved her bed downstairs. Later that evening she wanted to go to the toilet but collapsed on the floor. We called the doctor again and managed to get her back to bed. My wife and I looked after her at home for 7 days when a district nurse was provided. My mother had little control over her bowel or bladder functions and she was admitted to the local cottage hospital two weeks after the onset of the illness. A bladder catheter was inserted and she later developed pneumonia from which she did not recover for 6 weeks.

After 3 months, she was referred to a Rehabilitation Centre which was a long way from home. I felt she went there too soon. After 6 weeks she was discharged back to the local cottage hospital. I was told that she had made insufficient progress and that she was mentally

Introduction

confused. My mother stayed at the local hospital for a further 6 weeks during which time there was considerable pressure to take her home or put her into a nursing home. I managed to prevent her discharge, and she made quite marked progress being able to walk with a frame by the time of her discharge.

These cases illustrate what a stroke can mean to the patient and his family. It is now necessary to discuss what a stroke is.

2

What is a stroke?

The term 'stroke', although generally used by layman and doctor alike, lacks scientific precision because it covers a wide variety of different conditions and often means different things to different people. For example it is often considered to signify paralysis of one half of the body (hemiplegia) but, in fact, the pattern of disability following a stroke is extremely variable. There are also a number of terms used for stroke which need explanation, for example, cerebrovascular accident, cerebrovascular disease, transient ischaemic attack, cerebral haemorrhage, cerebral infarction. It may seem surprising that even doctors disagree about the definition of stroke, but the following definition has met general approval:

> A stroke is a sudden (acute) disturbance of brain (cerebral) function of vascular origin causing disability lasting more than, or death within, 24 hours.

There are certain key-words in this definition that need explaining. 'Acute' means that the onset or beginning of the illness is sudden but this does not mean within seconds or minutes, but perhaps hours rather than days. By 'disturbance of cerebral function' is meant brain damage and 'vascular origin' means a disorder affecting the vessels which supply blood to the brain as opposed to a tumour, which could produce a similar disability. If the vascular nature of the disorder is likely but not proven, then the diagnosis is a 'presumed stroke'. The 24-hour limit requirement is because if the disability is shorter-lived than this, then it is given another name (see below).

The medical term for 'stroke' is 'cerebrovascular accident', or CVA, but there may be confusion when the patient is described as having a 'left CVA' or a 'right CVA'. This is

What is a stroke?

because one side of the brain governs the opposite part of the body. It is not clear whether the doctor means that the patient with a 'left CVA' has an abnormality of the left side of the brain (left cerebral hemisphere) which will cause a right-sided paralysis (hemiplegia), or whether the term is describing a left hemiplegia, which would mean that the cause of the stroke was in the right hemisphere. Collecting details from the medical records of stroke patients (retrospective surveys) is fraught with this sort of problem and it is hoped that the term 'CVA' will eventually be dropped.

'Cerebrovascular disease' is a more general name used to cover a wide variety of conditions, including stroke.

Transient ischaemic attack

A *transient ischaemic attack* (TIA) is the name given to a stroke when the disability lasts less than 24 hours. 'Ischaemia' means insufficient blood supply. From what we know of the changes in the brain in stroke, if disability lasts more than 24 hours, then brain damage is more definite, whereas a TIA may not be associated with any obvious structural damage. There is an overlap between the brain changes occurring in stroke and TIA and this depends on a number of factors, for example the state of the other, non-damaged blood vessels supplying the brain and whether they can take over the circulation of the damaged vessels—in other words, whether the collateral circulation is adequate. The presence of high blood pressure (hypertension) is another factor. The differentiation of TIA from stroke is made on grounds of the description and duration of the illness. The distinction is important because the treatment of the two conditions is different: for TIA it is aimed towards the prevention of a more permanent (completed) stroke, whilst in the latter, treatment aims at lessening the degree of brain damage.

Strokes are divided into two main types: cerebral infarction and cerebral haemorrhage.

What is a stroke?

Cerebral infarction

The term 'cerebral infarction' includes the older terms of 'cerebral thrombosis' and 'cerebral embolus'. Cerebral infarction means that part of the brain is damaged due to lack of blood because the vessel feeding the area is blocked. In the past, the term 'cerebral embolus' was used for those cases in which an embolus (clot), suddenly caused a block in the blood vessel. It used to be assumed that the clot came from the left side of the heart, but recently the importance of hardening of the arteries in the neck (arteriosclerosis) has become recognized as a major source of emboli. The term 'cerebral thrombosis' was used for those cases where the stroke had a less sudden onset and the clot formed at the site of the blockage.

If the blood supply is not restored quickly the affected area of brain tissue dies (becomes 'infarcted'). As, in most instances, it is not possible to distinguish the cause of the blocked vessel, the term 'cerebral infarction' is now preferred to the older terms and includes both forms of 'thrombo-embolic' disease. Cerebral infarction can be defined medically as 'an area of brain in which the blood flow has fallen below the critical level necessary to maintain the tissue viability'.

Cerebral haemorrhage

Cerebral haemorrhage means a bleeding into the brain and is due to the bursting (rupture) of either a blood vessel or an aneurysm, which is a localized widening of the blood vessel, usually where it forks. In at least half the patients there is also a raised blood pressure (hypertension).

Other types of stroke

A patient with a more gradual paralysis, often developing over a period of several hours, rarely up to 3 days, is said to have an 'evolving stroke' or a 'stroke-in-evolution'. Although rare, it is important that it is recognized, as appropriate

What is a stroke?

treatment may prevent the development of the stroke. If the stroke does not get worse over an extended period of observation, the term used is 'completed stroke'.

'Reversible ischaemic neurological deficit' (RIND) is a term which has been suggested for the disorder of a group of stroke patients whose disability lasts more than 24 hours, but who make a full recovery, often within one week and always within one month. This type of stroke is separated from others because the long-term prognosis is probably better than for completed stroke and may even be better than for transient ischaemic attacks.

Strokes affecting one of the two halves of the brain (cerebral hemispheres) are five times more common than those affecting the brain-stem, which is that part of the brain that joins the hemispheres to the spinal cord. Transient ischaemic attacks also occur in the brain-stem through blockage of the vertebrobasilar system of blood vessels. In later life, usually over the age of 65 years, the blood supply to the brain-stem can be reduced by kinking or constriction of the vertebral arteries due to osteoarthritis in the neck, aggravated by movement of the neck. The legs may suddenly give way, making the patient fall to the ground without loss of consciousness (drop attacks) but these attacks of vertebrobasilar ischaemia need not result in permanent disability.

Why are all these different terms so important? First, all those concerned with stroke need to understand each other and so must speak the same language, that is, the type of stroke that has occurred must be explained or defined. Secondly, each of these types of stroke have a different course so that when a diagnosis has been made the outlook (prognosis) can be explained to the patient and relatives, and investigations and treatment can be planned. Thirdly, the follow-up information on stroke patients over a long period of time is relevant only if the types of stroke suffered are properly defined from the beginning.

What is a stroke?

How should a 'stroke' be described? To avoid confusion the part of the brain that is damaged as well as what caused it, and the degree of disability should be stated. As an example, a left cerebral hemisphere infarction causing a right hemiplegia leaves no room for doubt; the site of the lesion is in the left cerebral hemisphere, the nature of the lesion is cerebral infarction, and the resulting disability is a paralysis of the right arm and leg, i.e. a right hemiplegia.†

---

† A glossary of these and other terms is included at the back of this book.

3

How does a stroke occur?

In order to answer this question, we must start by knowing something of the blood supply to the brain and the way in which different parts of the brain work or function.

The blood vessels

Our knowledge of strokes goes back many centuries. Johann Jakob Wepfer, born in Switzerland, published in 1658 an account of four cases of apoplexy, which was the old term for stroke, and he confirmed by post-mortem examination that the cause was cerebral haemorrhage. He was able to dismiss the earlier teaching of Galen in the second century A.D., and demonstrate that the arteries at the base of the brain joined each other. He also declared that apoplexy could be caused when 'an influx of blood is prevented through the carotid or the vertebral arteries'.

The circle of arteries at the base of the brain was more fully described by Thomas Willis in 1664, since it has been called the circle of Willis. (Incidentally, Willis was also the first person to coin the term 'neurology'.)

The two main arteries supplying the cerebral hemispheres of the brain are the left and right carotid arteries. The ancient Greeks knew the effect of pressure on the carotid artery and it was Herophilus who suggested the name *carotid* (Greek *karas* = drowsy). In normal circumstances, the left carotid artery supplies the left cerebral hemisphere whilst the right carotid artery similarly supplies the right. Although their branches communicate, there is usually no flow through this joining vessel (anterior communicating artery (Fig. 1)). About 20 per cent of the blood leaving the heart goes to the brain. The rate of blood flow in the carotid arteries is 300 millilitres per minute.

How does a stroke occur?

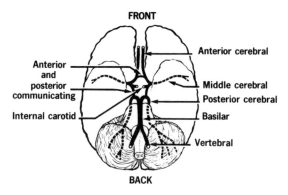

Fig. 1. The arteries at the base of the brain, which is shown as a section cut from front to back. The two vertebral arteries join to form the basilar artery. This vertebro-basilar system is connected to the internal carotid artery system by the posterior communicating arteries, which arise from the posterior cerebral arteries, to join the middle cerebral arteries. The circle thus formed is named after a famous English physician, Thomas Willis (1621–75) (Circle of Willis). The original picture was drawn by Sir Christopher Wren.

The brain-stem, cerebellum (the part of the brain that governs movements), and the back of the two cerebral hemispheres are supplied by two vertebral arteries which travel up in a special canal in the neck (cervical) vertebrae; it is in this area that these blood vessels can be squeezed by arthritic bony projections from the vertebrae. In most people, one vertebral artery is bigger than the other—a developmental or 'normal' variant. The blood flow in the vertebral arteries is less than in the carotid arteries: 100 ml per minute. The back of the cerebral hemispheres is supplied by the two posterior cerebral arteries which come off from the basilar artery, formed by the joining of the two vertebral arteries (Fig. 2).

Although the arteries at the base of the brain do communicate, circulation of the blood from one side of the brain to the other occurs only if there is a blockage in one artery, particularly if this is gradual as opposed to sudden. This transfer of blood from a well-supplied area to another of

How does a stroke occur?

reduced supply can be demonstrated by a special X-ray investigation called angiography (p. 57); this is important because the collateral circulation, if effective, can prevent a stroke.

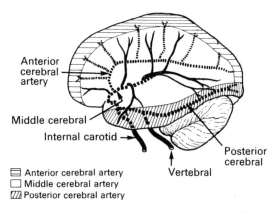

Fig. 2. Blood supply of brain (side-view)

The direct continuation of the internal carotid artery is the middle cerebral artery. The posterior cerebral artery on one side is linked to the middle cerebral artery by the posterior communicating artery. (Fig.1). Many small branches arise from each of these major arteries, perforating the brain and supplying vital areas. The border zone where two arterial beds come into proximity with each other is important because it is the area most likely to be affected if the blood flow in one or other of these arteries suddenly falls. Failure of the blood supply to an area of the brain, if not rapidly reversed, will cause death of an area of brain tissue (infarction), the commonest type of stroke.

Mapping brain functions

If the artery which has been blocked is known, the type of disability produced can be predicted since we know which

How does a stroke occur?

function of the brain will be affected. The doctor deduces from clinical examination the probable area of brain damage. 'Where is the damage, what is the damage, and why did it occur?' are the questions the doctor has to answer. To answer the question 'where is the damage?' we must discuss the 'new functional atlas of the brain' that has been demonstrated by many different techniques.

The ancient Greeks knew that the left hemisphere of the brain controlled the right side of the body and vice versa. In 1861 a French surgeon named Paul Broca described the findings of post-mortem examinations that he had carried out on two patients, both of whom had suffered from long-standing severe speech difficulties and paralysis of the right side of the body. Broca found that in both cases a particular area of the left hemisphere was damaged and concluded that this was the probable 'speech centre' of the brain.

With similar observations, others since have attempted to produce an atlas of the brain, relating body function to various parts of it. These early findings were often based on the study of small numbers of patients but this was to change during World War I when physicians were able to study many brain-damaged soldiers.

More specific mapping was achieved in the 1940s, particularly by Wilder Penfield of the Montreal Neurological Institute who, during certain neurosurgical operations, used small electrodes to stimulate electrically different areas of the brain. The early part of the operation leading to exposure of the brain through a surgically prepared skull flap (craniotomy), and the subsequent operation were carried out under deep anaesthesia so that the patient was unconscious and quite unaware of the procedure, but the electrical stimulation was carried out under light anaesthesia so that the patient could verbally respond. For example, as different nearby areas of brain were stimulated the patients would report that they felt their face twitch or that they could not find the right words for what they wanted to say. Animal experiments have also been used to confirm these observations.

How does a stroke occur?

These findings generated much excitement because this 'brain atlas' replaced the porcelain head used by the nineteenth century doctors who claimed to be able to deduce character from bumps on the skull (phrenology). In addition to the speech area, 'strips' of the brain were defined governing movement, sensation, and vision. Most interest was focused on the left hemisphere, known as the dominant hemisphere because of its language function, and for a time the right hemisphere was considered not to be so important.

Despite a greater understanding of brain function, there still remained large areas of the brain which appeared 'silent', particularly the frontal lobes. Surgical removal of a part of the right frontal lobe seemed to cause no physical disability although the patient's personality was sometimes changed. It was generally considered that these 'silent areas' were not truly silent but were concerned with other functions of which little was understood, for example, learning, memory, mood, and personality. Psychologists were particularly interested in trying to develop tests which could also detect disorders of these functions.

Observations based on patients with traumatic brain damage was known to have definite limitations. The method can be compared to firing a bullet at a television set and observing the internal damage and consequent poor functioning of the monitor, but this is hardly the best way to understand the complex interactions of the various components of the television to produce a picture. A new approach was needed.

In 1890 Charles S. Sherrington, working on animal experiments in Cambridge, noted that greatly increased brain activity, as occurs in an epileptic seizure, was associated with brain swelling due to greatly increased blood flow and, nearly a quarter of a century later, Barcroft, working in the same University, developed the hypothesis that a tissue can increase its functional level (work harder) only by increasing its rate of oxygen consumption. Since the brain's energy requirements depend on oxygen carried in the blood stream, the

How does a stroke occur?

functional activity of the brain can be measured by estimating cerebral oxygen utilization or cerebral blood flow. Another technique of measuring cerebral blood flow in animals is by means of thermocouples, whilst another, first done by Kety in 1944, is by analysis of the difference between arterial and venous blood samples after inhaling nitrous oxide.

The most exciting 'functional brain maps' yet seen were pioneered by Lassen, Ingvar, and Skinhøj at Bispebjerg Hospital, Copenhagen and the University of Lund, Sweden, in 1961. Their technique is to inject a small amount (bolus) of xenon-133, a radioactive isotope of the inert gas xenon, into one of the main arteries of the brain. The arrival and subsequent washout of the radioactivity from different brain regions is determined by a gamma-ray camera which has over 250 scintillation detectors, each scanning one square centimetre of brain surface. The information from this bank of detectors is relayed by means of a digital computer to produce a picture on a colour television monitor. Small squares of colour represent brain tissue activity, blue representing 20 per cent below normal activity and red 20 per cent increased activity. When the subject is asked to move his hand the corresponding hand-finger area in the central cortex 'lights up' and a supplementary motor area is also shown to be activated. This association area was a new discovery, not detected previously by any other method. Counting aloud showed areas of activity in the mouth area, the supplementary motor area, and the auditory area of the brain, whilst speaking activated three centres in *each* hemisphere—the motor and sensory parts of the mouth-tongue-larynx area, the supplementary motor cortex, and the auditory cortex. The areas in the right hemisphere are less clearly defined; the mouth-tongue-larynx and auditory areas appear to merge. Normal subjects and patients with different diseases such as stroke and schizophrenia have been studied. At present, this very new technique is not yet available for routine use but it is certain to increase our understanding of brain function in the years to come.

How does a stroke occur?

Different parts of the body are affected by a stroke depending on the areas of the brain that are damaged so that by examining the patient the doctor will be able to determine the probable damaged areas of the brain. A similar understanding by the patient and his family of the function of the different parts of the brain will help them to appreciate why the patient's problems are often so dissimilar to those of others who have also had strokes. In order to provide a background to the different types of stroke seen, some of these functional areas will be described.

Frontal area

Disorders involving the frontal areas are characterized by mood disturbances and personality changes. After a stroke, a patient may have difficulty in controlling his emotions, laughing or crying inappropriately. This can be particularly distressing for the patient who might, for example, burst into tears when he is in fact happy to see somebody, and for the family, particularly if the nature of the rapid mood fluctuations has not been adequately explained.

Both the left and right frontal areas have important centres for controlling the movement of the eyes towards the opposite side of the body, as well as centres for the sense of smell, which is located on the under (inferior) surface of the frontal area. The left frontal lobe also contains Broca's speech area as well as an area for writing which, if damaged, produces *agraphia*, an inability to write.

Speech area

The left hemisphere is said to be *dominant* for speech. Handedness is often used clinically to decide which is the dominant hemisphere. Ninety per cent of the population are right-handed and have their speech centre in the left hemisphere. Of the ten per cent who are left handed, the majority of these still have their speech centre in the left

How does a stroke occur?

hemisphere, but in the remainder the speech centre is in the right hemisphere.

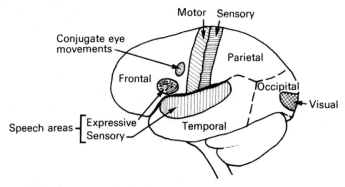

Fig. 3. Side view of brain. The lobes of the brain (frontal, parietal, temporal and occipital) are purely descriptive terms based on prominent dividing crevices (sulci) on the brain surface. Willed movement originates from the motor area. Speech, visual, sensory areas are also shown.

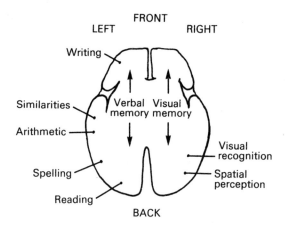

Fig. 4. Psychological assessment. The psychologist uses a battery of tests to assess the function of the frontal, temporal, and parietal lobes of each hemisphere. The results of these tests can be correlated with the physician's examination of the patient and with special tests, e.g. CT scan, and can be used together with other assessments to determine the patient's suitability for rehabilitation (after McFie 1979).

How does a stroke occur?

There are two speech areas in the left hemisphere: one of these is in the frontal lobe (Broca's area) and is involved in the initiation of speech. Damage of this area will cause the patient to have difficulty in finding words. Typically the patient searches for the correct word or words and his sentences are short, poorly formed, and lacking fluency. This speech disorder is termed *expressive dysphasia*, or difficulty in the symbolic formulation of words (Fig. 3).

The second (Wernicke's) speech area is further back (posterior) in the temporal lobe and is involved in the monitoring and reception of speech. Damage here causes a severe speech problem so that the patient is unable to understand what is being said to him, although his hearing is intact, and he is unable to monitor his own speech for mistakes. He makes frequent errors and speaks nonsensical words. Despite this, fluency is retained. A conversation with such a patient is often likened to that experienced at cocktail parties—namely, each person provides a dialogue without any real communication occurring between them. This disorder is termed a *receptive dysphasia*.

It is uncommon for a patient to have a pure expressive or receptive dysphasia and it is more usual to find a mixed pattern, often with one type predominating.

Motor area

The 'fissure of Rolando' is a prominent groove (sulcus) on the surface of the brain. In front of this fissure lies the motor area, where there are brain cells which control movements of the *opposite* side of the body. The body areas are represented as an 'upside-down man', the leg area being at the top and the face area towards the bottom of the Rolandic fissure. A disproportionately large area is devoted to the hand, face, and tongue, because of the importance of speech and hand function development in *Homo sapiens*, as opposed to other species.

How does a stroke occur?

Sensory area

The sensory representation in the cortex is again an 'upside-down' man, but not exactly matching that of the motor area. It is situated behind the fissure of Rolando and sensations, whether conveying touch or pain, are relayed ultimately to this area of the brain for interpretation. Because it also receives the sensations of the position of joints when we move an arm or leg, it also helps as a 'controller' of movement.

Temporal area

The left temporal lobe, which contains Wernicke's speech area, is also concerned with 'verbal memory' and reading. The term used for reading difficulties is dyslexia (Fig. 4).

Parietal area

Damage to the left parietal area affects calculation, for example adding up and subtracting, which can be tested by simple arithmetic. A disturbance of this function is called acalculia. Constructional ability is also affected and this can be tested by asking the patient to draw, for example, a house, flower, or clock face. There may also be difficulty in dressing or finding the way through a book.

The right parietal area is concerned particularly with perception and if a stroke affects this region the patient is totally unaware of his left side so that, when carrying out a bimanual task, he ignores the left hand as though it does not belong to him. When the damage is further back towards the occipital area, the patient may have *visual inattention*, i.e. ignoring the left field of vision. As a result, if he reads, he does not see the left half of the page.

Both *sensory* and *visual* inattention can be particularly incapacitating, even though the nature of the problem may

How does a stroke occur?

not be immediately obvious, since there is no paralysis as the motor area is not involved.

Visual area

For each eye there is an outer (temporal) visual field and an inner (nasal) visual field. The visual area is situated in the occipital lobe, the right visual area receiving information from the left fields of both eyes (the nasal field of the right eye and the temporal field of the left eye) (Fig. 5). This is because the nerve fibres transmitting impulses from the nasal half of the left retina (which serves the left temporal field) crosses over. Damage to the right occipital region or to the nerve pathway leading to it (the optic radiation) results in loss of vision towards the left side. Since the same side of each eye will be affected (homonymous), each losing one half of the visual field (hemianopia) the difficulty is called a homonymous hemianopia.

The functional areas of the brain described do not work in isolation, but in association with each other, and the left and right cerebral hemispheres are interconnected by millions of nerve fibres. This process of integration of all brain areas is essential for our everyday activities.

This interrelationship has been graphically demonstrated by recent cerebral blood flow studies (p. 15). Even when the left hemisphere is involved in a task previously thought to be exclusively its own, such as speech, these studies have shown activity in the corresponding areas in the right cerebral hemisphere. They have also shown that, after a stroke, areas in the cerebral hemisphere *opposite* to the severely damaged one show reduced activity and this explains why, when some stroke patients perform psychological tests, abnormalities of both hemispheres are detected.

Motor neurones

There are two types of nerve cells concerned with movement

How does a stroke occur?

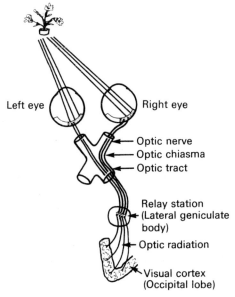

Fig. 5. Visual fields. Visual information from the left visual field is relayed to the visual cortex in the right occipital lobe. A stroke which damages brain tissue in the right cerebral hemisphere will cause a left hemiplegia, and if the right optic radiation is affected, a left homonymous hemianopia (homonymous = same visual defect in both eyes; hemianopia = half of the visual field of each eye; left = side of the visual field affected).

(motor neurones), upper and lower. The nerve cells in the grey matter (cortex) of the brain controlling voluntary movement send out long nerve fibres (axons) which run through the brain-stem to end at various levels in the spinal cord. These long axons are called upper motor neurones and, in their course in the brain, they run through an important narrow area called the internal capsule where they lie in close proximity to sensory fibres. The area is supplied by a branch from the middle cerebral artery (Fig. 2). The lower motor neurone is linked to the upper motor neurone by a junction (synapse); it begins in the spinal cord and runs out

How does a stroke occur?

in the peripheral nerve to make muslces contract. Paralysis or weakness of the opposite side of the body will be caused by interruption of the upper motor neurone at any point in the brain, the common places for this to occur being in the internal capsule of the cerebral hemisphere or the brain-stem.

Electrical impulses run out along motor neurones and return to the brain via sensory neurones, whilst other fibres link the brain-stem to the cerebellum and higher centres. In the brain-stem are also the nuclei of the lower motor neurones supplying the head, for example, the eye muscles and muscles of the face and tongue. Where an upper motor neurone to the face is damaged by a stroke, paralysis of the facial muscles is incomplete and affects only the lower half of the face.

The cerebellum is concerned with the coordination of fine movement so that infarction or haemorrhage here leads to incoordination of the limbs. This is less common than strokes affecting the cerebral hemisphere.

Having discussed where the damage in the brain can occur and some of the known functions of these brain areas, we should try to answer the question 'why do strokes occur?' The answer lies partly in the associated relationship with hardening of the arteries (arteriosclerosis) and high blood pressure (hypertension).

Arteriosclerosis

As we get older, everyone develops hardening of the arteries, often likened to the 'fur' in a kettle or a lead pipe. This 'furring' process is due to the development of *atheroma*, which literally means porridge, and refers to the gruel-like properties of the 'fur'. This process is referred to as atherosclerosis or arteriosclerosis and can lead to almost complete blockage (occlusion) of the blood vessel. Atheroma differs from the 'fur' in the kettle or lead pipe in that it cannot be scraped off since it is part of the vessel wall.

How does a stroke occur?

A new-born infant has no arterial atheroma but, as early as the age of six months, minimal atheroma may already be present. The development of atheroma is related to the development of high blood pressure (hypertension) but which comes first is uncertain. Patients with hypertension certainly have an accelerated type of arteriosclerosis. On the other hand, some families have been described who have an inherited fat (lipid) abnormality and develop widespread atheroma at an early age without the development of hypertension.

One of these lipids from the blood stream (cholesterol) is incorporated into the vessel wall and an atheromatous plaque is formed, which has the appearance of a yellow streak. These plaques first develop in the arch of the aorta (the main artery from the heart from which all the other arteries branch), usually causing little trouble. The plaque is a roughened area on the vessel wall and various constituents of the blood, such as red blood cells, platelets, fibrin, and lipids may stick to the plaque and may be incorporated into the vessel wall. In this way, atheromatous plaques will increase in size over the years. Atheroma tends to collect at the points where arteries branch (bifurcations), i.e. junctions where there is often turbulent blood flow. Its predilection for certain sites such as the coronary arteries or the bifurcation of the common carotid artery is well known.

These plaques can act as a source of emboli: i.e. platelets, fibrin, or cholesterol can be released from the surface of the plaque, travel in the blood stream and subsequently block a major vessel. This block may be temporary, as in a transient ischaemic attack, or permanent, when cerebral infarction will occur. Alternatively, if the area of atheroma is extensive, the diameter of the artery (lumen) becomes progressively smaller (stenosed). Blood flow is not affected until the arterial lumen is almost completely occluded. To remove an area of atheroma from an artery the surgeon has also to remove the inner arterial layer or intima, a procedure known as endarterectomy. Atheroma also develops at the site of tears in the inner

How does a stroke occur?

lining (intima) of the arterial wall. These tears may be related to hypertension. If we know those factors that influence the development of atheroma, then prevention of strokes should be conceivable (Chapter 4).

Hypertension

The relationship between atheroma and hypertension can be likened to the 'chicken and egg' argument, since the presence of hypertension will predispose to arteriosclerosis and vice versa. The patients at risk from cerebral haemorrhage often have an accelerated form of hypertension, particularly if poorly controlled by treatment. There are still problems to be resolved concerning the normal levels of blood pressure and how to measure the blood pressure, on which subject, surprisingly, opinion remains divided, but the recognition and treatment of hypertension is a very important aspect of preventive medicine. Why a person should suddenly become hypertensive is not known but a great deal of research has gone into this topic.

4

Risk factors

In the previous chapter, we described the development of arterial disease that occurs, to a greater or lesser extent, in all of us as we get older. Many comparisons have been made between the two major types of vascular disease, ischaemic heart disease and stroke. Although they have been considered as fundamentally similar, because of the same underlying pathology (atherosclerosis), there are major differences between the two diseases, not least the age of occurrence. Recent population studies in the country and the United States have shown a significant decline in stroke, particularly cerebral haemorrhage, as a cause of death. There has not been a concomittant fall in the mortality rates for ischaemic heart disease over the same period of time but recent trends suggest that the mortality rate is now beginning to fall.

Table 1. *Annual incidence of stroke related to age*

Age group	Incidence per 1000 per year
35–44	0.25
45–54	1.00
55–64	3.50
65–74	9.00
75–84	20.00
85+	40.00

Strokes can occur at any age, from newborn babies to octogenarians. Table 1 shows how the incidence of strokes rises rapidly with age. Incidence refers to the number of new cases that occur in a defined population in a given

Risk factors

period, e.g. per 1000 in 1 year. The average incidence rate for stroke in most countries is slightly more than 2 per 1000 population per year, although Japan and Finland report higher figures. This means that in an area served by a district general hospital where the catchment is usually 250 000 people, there will be at least 500 new cases of stroke per year, and that a general practitioner, who usually has an average of about 2500 people on his list, will see about 5 new cases per year. Only half of all the new stroke cases are admitted to hospital and of those, only a small percentage will be admitted within the first 24 hours. One-third of all patients surviving the acute illness have persisting major disability, and it is estimated that at any one time in England and Wales, 10 000 hospital beds are occupied by stroke patients. Stroke is the leading cause of disability in the Western world, and the socio-economic implications are enormous. A recent international multicentre study was carried out under the auspices of the World Health Organization to obtain accurate information on stroke morbidity and mortality in different parts of the world by means of stroke registers. Seventeen centres from 12 different countries participated, and it was estimated that the total number of new stroke cases occurring every year in Europe alone was in the order of 1 million, accounting for at least 30 million hospital days.

Risk factors and prevention

The profile of a potential heart attack victim has been popularized in recent years in order to make people more 'health conscious'. The person most likely to get a heart attack is a man, about 50 years old, probably in a managerial or supervisory post, who travels to work by car. His job will be typically stressful; he over-eats, takes no exercise, and smokes more than 20 cigarettes a day. His blood pressure, if checked, is likely to be at the upper limit of normal or, more probably, found to be significantly raised. The risk

Risk factors

factors illustrated are (1) age; (2) sex; (3) obesity; (4) smoking; (5) raised blood pressure; and (6) a stressful life-style. To this must be added hereditary and familial factors or predisposition. How does this compare with the risk factors for stroke? (The following are not nessarily listed in order of importance.)

Age

The importance of age when considering stroke incidence has already been mentioned (Table 1). The mortality from stroke doubles with each decade after the age of 40 years and by the age of 80 years, 1 in 3 people will have suffered some type of stroke illness.

Sex

Stroke tends to affect men at younger ages; women at older ages. The protective effect of female hormones in preventing arterial disease until the menopause is thought to be a factor that explains the predominance of ischaemic heart disease in men since, after the female menopause, incidence rates in women begin to approach male rates.

Contraceptive pill

The contraceptive (oestrogen) pill is thought to increase the risk of stroke amongst young women. However, some studies in Rochester, Minnesota among all female residents failed to demonstrate any change in the stroke rate between the years before the introduction of oral contraceptives and after; cerebral infarction was commoner in women in the older age group, 70–79 years, who had used oestrogen pills for menopausal and other symptoms, but this was probably due to the effect of oestrogens in raising the blood pressure. Oestrogens are also given to men with cancer of the prostate gland and an increased risk of stroke has been found in these patients too.

Risk factors

Heredity

Since we cannot choose our parents, one factor that none of us can control is our genetic inheritance. Because some families are more prone to vascular diseases (heart disease, stroke, hypertension) than others, the family history is always important.

There are also racial differences, for example in the United States, black people are more susceptible to vascular disease than whites. In India, strokes are thought to be commoner in younger people than in the Western world, but this might be explained by the lower expected age of survival in the population at large. In Japan, incidence rates for stroke and, in particular, cerebral haemorrhage, have been consistently reported far in excess of those in other countries and the World Health Organization study also found differences when Japan was compared with Europe. Interestingly enough, Japanese living in America have a lower incidence than Japanese living in Japan. Stroke patients in Japan were more frequently in coma on admission (40 per cent as opposed 25 per cent of European patients), stayed in hospital longer (ratio 87 per cent:50 per cent at 3 weeks), and fewer of the patients received rehabilitation (ratio 50 per cent:25 per cent). Hypertension was also commoner in the Japanese patients. One year after the stroke, nearly all the Japanese survivors were at home, whereas 25 per cent of the European survivors were in hospital or nursing homes. Although it is possible that differences in incidence rates between Japan and the rest of the world are artefactual, they could be genuine, in which case further study of these differences may give us further clues as to risk factors.

Socio-economic status

Stroke incidence rates are highest among those people in the lower socio-economic groups. In Great Britain, there is a geographical variation, with the north and west of the country

Risk factors

having mortality ratios above, and the south and east below, the national average. The pattern is similar for heart disease.

Season of the year

Strokes probably occur more commonly during the winter months.

Hypertension

This is the single most important risk factor. The relationship between hypertension and stroke has been known for a long time and it is likely that the falling incidence rates for stroke is at least partly due to the improved recognition and treatment of raised blood pressure.

Lipids

There is a strong association between raised levels of lipids (cholesterol and triglycerides) in the blood (hyperlipidaemia) and the development of atheroma in the heart's (coronary) blood vessels. In the case of stroke such a relationship has been difficult to prove since hyperlipidaemia is often present in patients who also have hypertension. Atheroma is less common in the cerebral arteries than in the coronary arteries. The lipid abnormality has to be present for many years before cerebral atheroma develops so that most of these patients would be more at risk from dying of atheromatous coronary artery disease than cerebral artery disease. Blood lipid levels, in any event, tend to rise with age, so that no definite relationship has been proven in stroke patients.

Diabetes mellitus

This disease is commoner in stroke patients than in a normal population of similar age. Since there is an association between diabetes, hyperlipidaemia, and hypertension, it is difficult to be certain of a relationship between diabetes and stroke when

Risk factors

these other conditions co-exist. Diabetes probably does increase the stroke risk but when the three conditions exist in the same patient, then the relative risk is proportionally much greater.

Heart disease

There is a definitely increased risk of stroke in patients with heart disease (whether it be chronic disease, acute heart attacks, or implanted (prosthetic) heart valves). Sudden disorders of the heart rhythm, which produce cerebral infarction due to an embolus, enlargement of the left ventricle of the heart, and abnormalities seen on the electrocardiogram (page 53) are all associated with an increased risk of stroke.

Transient ischaemic attacks (TIA)

About 30 per cent of all patients who have had a TIA are at risk of having a stroke within 2 years, half of these occurring in the first year after the first attack. The risk is greatest in the first month, becoming less after the first 3 months.

Obesity

Although there is a relationship with coronary artery disease, there is no definite evidence to suggest a relationship between stroke and obesity.

Cigarette smoking

Again, there is a strong relationship with heart disease but little evidence to link cigarette smoking with stroke.

Alcohol

Alcohol is not thought to be a risk factor, and drinking in moderation may on the contrary even be beneficial!

Risk factors

Exercise

The need for exercise in the prevention of heart disease has been much debated in the medical and lay press, and there has been a meteoric rise in the popularity of 'jogging' so that there is even a monthly 'jogging' journal for enthusiasts! However it is not possible to make any categorical statements about whether lack of exercise is a risk factor.

Other factors

Many other diseases can be mentioned which may cause or lead to stroke, e.g. sickle cell disease (page 47), blood diseases associated with increased blood viscosity, and syphilis. Diagnosis depends on adequate investigations after which the appropriate treatment can be given. In any person who has just returned from overseas and develops acute neurological symptoms, the possibility of cerebral malaria exists.

The importance of risk factors causing stroke is that removing them could lead to effective prevention. Because of the problems of treatment after stroke has occurred, prevention is probably the most important area of stroke research and development. Despite the large number of risk factors suggested, a definite association has been proven only with hypertension, heart disease, and transient ischaemic attacks. Diabetes is another probable factor, especially if associated with hypertension or hyperlipidaemia. Of these risk factors, detection and treatment of hypertension in the population is the single most important stratagem in the prevention of stroke. Patients who have had a transient ischaemic attack should always consider this as a warning sign and be referred to the appropriate specialist for assessment, investigations, and treatment, either medical or surgical. Unfortunately, there is no evidence that good control of diabetes prevents strokes, nor that any drastic change in diet, cigarette smoking, or alcohol consumption will prevent them.

Risk factors

As heart disease is the most important cause of death in patients who have survived their stroke, it is wise for obese patients to reduce their weight and for patients to give up cigarette smoking.

Strokes in children

Although the incidence rates of stroke in children are not known for certain, the list of the different types of stroke (Chapter 5) is even longer than in adults; hemiplegia in children is not very rare but the cause is often unknown. Subarachnoid haemorrhage (p. 44) is very rare in children but is more likely to be due to an angioma than an aneurysm. Sometimes there are congenital abnormalities of the cerebral blood vessels which may lead to areas of brain infarction either *in utero* or in early life. Brain damage can also occur during a difficult protracted labour. Another cause of stroke in children is the so-called 'pencil injury' caused by a child falling over whilst having in his mouth a pointed pencil, which penetrates the tonsillar fossa, to damage the carotid artery in the neck. Infection can also spread through this route to cause an inflammation of the carotid artery—an arteritis. Sudden occlusion of the carotid artery can lead to cerebral infarction with paralysis of the opposite side of the body.

In developing countries, malnutrition is common and predisposes to infection. In a child with sickle cell disease (p. 47) this can provoke a crisis, and about 4 per cent of children can develop a stroke during this acute episode. In Nigeria, Professor Osuntokun has found that 10 per cent of all stroke cases are under the age of 30 years, most with no obvious cause. Arteritis is a possibility and malaria affecting the brain is another. Meningitis, due either to the pneumococcus or the tubercle bacillus, can cause hemiplegia in 10 per cent of cases. Improved maternal, neonatal, and child health care in the years to come will materially improve this dreadful toll of young lives.

5

The different types of strokes

The way in which a stroke starts often gives a clue as to its cause but, without special investigations, there is only a 60 per cent chance of getting the whole answer correct. The following case histories illustrate some of the ways in which different types of stroke begin and their differing causes.

Stroke causing unconsciousness

Case 5. A 62-year-old-man stopped taking his tablets for high blood pressure after 15 years. Eighteen months later, he tried to push his car after it had refused to start on a cold morning. Shortly afterwards, he complained of a headache and one hour later vomited. He went to lie down and, 3 hours later, his wife could not wake him. She immediately called the family practitioner who arranged hospital admission. In the casualty department, the patient was found to be deeply unconscious, not responding at all. On moving the limbs passively, the casualty officer found that all four limbs were floppy (flaccid). Over the next few hours, he improved slightly in that his left arm and leg moved in response to painful stimuli, but he died the next day.

This story with its very sudden onset, headache, vomiting, and rapid loss of consciousness is typical of a bleed into the brain (cerebral haemorrhage). Other pointers were the history of hypertension, particularly the recent period off treatment, when the blood pressure could have risen unchecked, and the relationship with heavy, sudden exercise. The initial paralysis was severe and there was no difference on the two sides suggesting a massive bleed. The clinical diagnosis of a left cerebral haemorrhage was confirmed at a post-mortem examination.

The mortality from cerebral haemorrhage is very high, exceeding 80 per cent. If the patient survives the initial episode or if the haemorrhage is small rather than 'massive',

The different types of stroke

the degree of residual neurological deficit may, surprisingly, be less than that which occurs following cerebral infarction. The reason for this is that the blood, coming from a small ruptured artery, does not destroy the substance of the brain but separates the nerve fibres, so that when this blood is absorbed, the nerve fibres remain intact and function well. This is in contrast to cerebral infarction, where some brain tissue is severely damaged so that it does not recover.

Stroke affecting the left side of the brain

Case 6. A 74-year-old woman awoke one morning and found that her right leg and foot felt heavy. During the day, she frequently caught her right foot on the carpet whilst, in the afternoon, she tried to write a letter and found that she could not use her right hand. She went to bed early; during the night, she tried to get out of bed to go to the lavatory and found that she could not move her right arm and leg properly. Her husband waited until the morning before calling the general practitioner and she was then admitted to hospital. There, she was fully conscious. Her speech was affected in that, although appearing to understand, she could not find the correct words for simple objects and had great difficulty in constructing simple sentences. When making these verbal mistakes, she became frustrated and upset with herself. The right arm and leg were flaccid and when asked to move her right arm she either shrugged and shook her head or attempted to lift the right arm with her non-affected, left hand. It was noticed that the right side of her face was also affected, the angle of the mouth being drawn down and the normal line running down the face from the side of the nose to above the angle of the mouth (nasolabial fold) was flattened. Her eyes were deviated towards the left side and she could not see to the right (owing to a defect in the visual fields). She also had less feeling to touch and pin-prick down the affected side. During the next 24 hours, she became drowsy but would open her eyes to command. She was incontinent of urine on several occasions. The paralysis of her right side remained the same but, after 6 days, the tone had increased and these limbs became stiff (spastic). Her drowsiness persisted for 4 more days after which she improved. The speech was still affected and, although her eyes were no longer deviated to the left, the visual field defect still remained.

This woman's stroke was due to damage (infarction) to the left side of the brain (cerebral hemisphere), affecting those

The different types of stroke

areas governing speech, movement, and sensation. In addition, the visual pathway (optic radiation) had been damaged, causing the loss of vision towards the right side, i.e. the right outer or temporal half of vision of the right eye and the right inner or nasal half of vision of the left eye (a right homonymous hemianopia).

The gradual onset, with the story of increasing weakness, indicated that the cerebral infarction was due to a thrombosis or blockage of an already diseased, atheromatous blood vessel. The extent of the damage, involving a large area of brain, indicates that the blockage was probably in the left internal carotid artery. The gradual onset could be explained by the 'flow-failure'; an area of brain was deprived of its major blood supply. As blood flow decreased below the critical level, the damage to nerve tissue in this area became permanent.

Fall in the level of consciousness (drowsiness or coma) is a very important sign which usually indicates the presence of swelling of the brain (cerebral oedema) (Chapter 6). The eyes are deviated because the corresponding gaze centre in the frontal lobes is not working and the unopposed action of the normal gaze centre on the other side of the brain moves the eyes to the opposite side, i.e. towards the side of the damaged hemisphere and away from the paralysed limbs.

These three signs: low level of consciousness, the eyes deviated away from the paralysed limbs, and the presence of a flaccid hemiplegia usually indicate a high early mortality from cerebral infarction.

A patient with a normal level of consciousness throughout the first 48 hours has a very good chance of survival, approaching 85 per cent. If the level of consciousness is severely reduced and the patient remains unconscious, not reacting at all during the first 48 hours, then death is almost inevitable.

Because cerebral oedema is the cause of early death from cerebral infarction, treatment with drugs is directed towards this, recovery of consciousness and voluntary eye movements being good signs of recovery.

After the first 4 days, cerebral oedema is gradually absorbed

The different types of stroke

and, for this reason, the critical stage is often in the first week following the stroke. After this point, assessment of the amount of brain damage allows a prognosis to be made concerning the likely degree of immediate recovery and the eventual outcome in terms of functional ability.

Stroke following a heart attack

Case 7. A 66-year-old man suddenly complained, after a day's outing, of left-sided chest pain going down his left arm, which lasted 40 minutes and was not relieved by his usual tablets. Having been seen by his general practitioner who diagnosed a heart attack (coronary thrombosis), he rested in bed for the next 3 days. He had retired prematurely because for the previous 3 years he had suffered from attacks of angina which were brought on by exercise and usually lasted only 2 minutes, being relieved by tablets sucked under the tongue.† On this occasion the tablets did not work indicating that the nature of the pain and its duration was different to preceding attacks. On the fourth day after his heart attack, he suddenly developed paralysis down the right half of the body (right hemiplegia).

This patient sustained a left cerebral hemisphere infarction following a heart attack (myocardial infarction) due to obstruction of the arteries feeding the heart (coronary thrombosis). Over the damaged area of heart tissue, a clot (mural thrombus) forms from the blood elements, designed to protect the damaged area and aid the healing process. As part of the healing process, the clot is broken up (lysed) between the fourth and seventh day; particles are removed in the blood stream and can be a potential danger elsewhere in the body (embolus).

After a myocardial infarction, the left ventricle is most commonly damaged and a clot formed here can be swept

---

† These were glyceryl trinitrate which dilates blood vessels and acts by temporarily reducing the amount of blood returning to the heart and hence the mechanical work the heart has to do. This mode of action explains its main side-effect, that of low blood pressure (hypotension), particularly if too many tablets are taken in a short period of time. Trinitrate is best taken *before* performing any activity likely or known to provoke an attack of angina.

The different types of stroke

along in the blood from the ventricle into the aorta, and from there into the blood stream of the carotid artery leading to the brain. Since the carotid circulation is in direct communication with the aorta and arises before the other main arteries, 50 per cent of all emboli originating from the heart block cerebral vessels. The left common carotid artery is in direct line with the aorta and this may explain why emboli more commonly affect the left rather than the right cerebral hemisphere.

Because of the high incidence of coronary-artery disease in patients with stroke (see Chapter 4), abnormalities of the electrocardiogram, for example abnormal rhythms or extra beats are common, and, for this reason, all patients with a stroke must be screened in case they have had a silent myocardial infarction. Five per cent of all patients may show evidence of a fresh myocardial infarction, which obviously raises the possibility that this may have been the cause of the stroke. Since atheroma is very common in the neck (carotid) vessels, this could be the cause of the embolus, particularly if investigations (Chapter 6) show there to be a significant narrowing (stenosis) of the carotid artery. It is difficult to predict which coronary patients run the risk of embolus but treatment of the heart attack could possibly protect patients from a subsequent stroke (Chapter 4).

Stroke affecting the right side of the brain

Case 8. A 57-year-old male airline executive had been working on a new air schedule involving flights to the Scilly Isles. The day before hospital admission, he had signed a cheque upside down and found that simple tasks, such as dressing, were impossible without assistance. The onset of his illness was sudden and he could not understand what was happening to him. In hospital he was bewildered and bemused and could not shave very well, particularly the left side of his face. He did not seem to be aware of his left side and tended to ignore objects placed on this side. He could not find the first page of a book and, when reading a page from the book, he missed out the first few words and never seemed to read from the left hand margin. He was asked to draw a rough outline of England and Scotland but, in spite of his previous skills

The different types of stroke

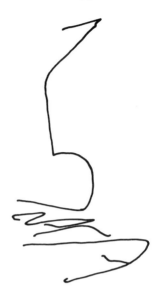

Fig. 6. Case 8 (see text) was asked to draw a map of Britain. In spite of having been an airline navigator, he has ignored the whole left coast.

Fig. 7. Case 8 was also asked to draw a bicycle. He omitted the rear wheel (in the left visual field).

The different types of stroke

and the fact that he had been an aeroplane navigator during World War II, his effort was surprisingly poor (Fig. 6). He had missed out all the left hand coast of Scotland, England, and Wales, and, significantly, the Scilly Isles. He could not site well-known cities such as Bristol. When asked to draw a house, the left side of the house, and the left side of the window frames were absent (see also Fig. 7). There was no impairment of vision, and motor and sensory testing of his left arm and leg was normal. There was some recovery but not enough for him to return to his job.

This patient had suffered a right cerebral infarction localized to one part of the brain concerned with constructional tasks (parietal lobe) (see Plate 1). The neglect of the left side accounted for his disabilities, including his difficulty in dressing. Vision was not affected in his case, but it can sometimes be affected, as demonstrated by the next case.

Case 9. A 51-year-old male clerk did not feel very well while at work during the day. He had no specific complaint but thought he ought to visit his general practitioner on the way home. Without any specific complaint, it is perhaps not surprising that no obvious abnormality was found. Driving home from the doctor's surgery, he hit three parked cars on the left-hand side of the road. Alarmed by this experience he telephoned his general practitioner and was promptly referred to the neurology outpatient clinic the next day (minus his car!). On examination, the patient did not have a left homonymous hemianopia, which could have explained the patient's mishap. However, when presented with visual stimuli in both temporal fields *at the same time*, he persistently ignored the object in the left visual field although this was recognized if presented singly.

This left visual inattention was due to a cerebral hemisphere infarction, again in the parietal area, but slightly further back in the brain than in the previous case. This patient had no previous history, but was found to have high blood levels of cholesterol. He recovered completely from his initial complaint within three weeks but had a myocardial infarction two years later, despite diet and cholesterol-lowering drug therapy.

The different types of stroke

Cerebral infarction (stroke in evolution)

Case 10. A 55-year-old woman was admitted to hospital following the acute onset of paralysis during a three-hour period on a Saturday afternoon. Her right arm and leg were affected but the paralysis was mild and the patient could raise her arm and leg although slight pressure from the examiner's hand on the outstretched limb was sufficient to overcome her efforts. She had a mild speech disturbance for naming specific objects (nominal dysphasia) without any understanding (receptive) problems. There was no alteration in the level of consciousness. Four hours after admission, she became more drowsy but was still easily rousable. Over the next 24 hours, the degree of paralysis in the right arm and leg increased and by 48 hours the patient had a flaccid hemiplegia with sensory loss over the right side. In addition, the patient's eyes were deviated towards her left side and she had an obvious right homonymous hemianopia (see page 21). She became more drowsy, the extent of her disability being greatest at 72 hours. She began to improve during the following three days with recovery of the eye movements, visual field loss and speech function, but her right side remained flaccid.

This patient had suffered a left cerebral hemisphere infarction and an arteriogram showed a complete block of the left internal carotid artery (see Plate 2).

An important sign on admission was of a harsh 'slushing' noise (bruit) heard with the stethoscope over the left side of the neck which had disappeared by the second day. The presence of this noise indicates turbulent blood flow caused by a major degree of stenosis of the left internal carotid artery.

This type of stroke with its slow progression is called a 'stroke-in-evolution'; although uncommon, it raises the hope of its progress being arrested. At the start of the stroke the left cerebral hemisphere dysfunction may have been due to a reduction in the blood supply due to blockage from an embolus coming from an area of atheroma in the left internal carotid artery. The subsequent loss of the bruit showed that there was no blood flow in the left internal carotid artery because it had become completely blocked.

The different types of stroke

Stroke of the base of the brain (brain-stem infarction)

Case 11. A 67-year-old woman suddenly felt dizzy whilst standing at the kitchen sink; the room seemed to go round and round and this sensation (vertigo) persisted for an hour. She felt nauseated, but did not vomit. Three hours later, the vertigo returned and she had great difficulty in walking; it felt as if her left side was not properly coordinated. Her husband brought her to the hospital casualty department, where she was fully conscious but still complaining of vertigo. Her speech was slightly slurred (dysarthria) although its content was quite normal. Although her vision was unimpaired, eye movements were noted to be abnormal, the eyes slowly moving back to a central position before quickly flicking back again, a sign known as nystagmus.† She had weakness of the lower half of the right side of the face but there was no weakness in the left arm or leg, although tone was slightly reduced (hypotonic). She found it was impossible to perform rapid movements of the hand and, when trying to touch her nose with her left forefinger, the incoordination of the left hand was obvious. Similarly, she could not run her left heel down her right leg and, when she tried to stand, she had difficulty balancing and could only walk with assistance.

The presence of vertical and horizontal nystagmus, right facial weakness and incoordination of the left arm and leg indicated that the stroke had affected the brain-stem. A haemorrhage into this area is rare, accompanied by rapid loss of consciousness and usually fatal. In patients surviving a brain-stem stroke, recovery is often very good.

These different types of stroke are by no means easily diagnosed and there are several other conditions that simulate stroke.

Transient ischaemic attack

If a stroke is of short duration, lasting less than 24 hours, with complete recovery, this is called a transient ischaemic attack, or TIA.

† This sign is found in a variety of neurological and ear conditions, being due to defective vision, or an imbalance of the balance organ in the inner ear (vestibular apparatus) or its connections with centres in the brain-stem.

The different types of stroke

Case 12. A 60-year-old male teacher suddenly developed loss of vision in the right eye, which he described as 'like a curtain coming down over my eye'. After one minute, the 'curtain lifted' and normal vision was completely restored. The following day he had three similar episodes, each followed by complete recovery. Two weeks later he went to his general practitioner because he had an episode of weakness of his left hand lasting 10 minutes. He described the episodes of blurred vision only on direct questioning, because he did not realize that they and the hand weakness had a common cause. In hospital, investigation showed a narrowed right internal carotid artery. After operation on this artery, he made a complete recovery, and remains well.

The pattern of vision loss is typical of a blockage of the artery to the back of the eye (retina). This is due to an embolus arising from an area of atheroma which caused the narrowing of the internal carotid artery. The typical attacks of sudden blindness accompanied by a carotid artery bruit confirm the diagnosis. Further investigations (arteriography; see Chapter 6) show the exact area of disease and indicate whether treatment should be medical or surgical.

Vertebrobasilar ischaemia (or insufficiency)

Case 13. A 75-year-old woman was stretching up to a top shelf in her kitchen when she suddenly fell to the ground, but after a few minutes recovered completely. On another occasion, whilst about to cross the road she turned her head from left to right and was saved from falling by a friend who was with her. When later describing this event, she described a feeling of sudden rotation.

These attacks are not uncommon in the elderly and are due to interruption of the blood supply to the brain-stem. The brain-stem is supplied by the vertebral arteries which are of unequal size, one artery usually providing most of the blood. These arteries travel to the brain-stem up a bony channel (cervical canal) in the spine of the neck (cervical vertebrae). In older people, osteoarthritis of the cervical bones develops, causing narrowing of the cervical canal, particularly when the head is extended or rotated.

The different types of stroke

Case 14. A 62-year-old vicar was showing an architect the structural faults of his church steeple. Suddenly pointing with his left arm to illustrate his comments, he fell to the ground unconscious, recovering completely after 4 minutes.

In addition to the mechanical problems described above from the wear and tear in the cervical vertebrae, this patient had a stenonis of the first part of his left subclavian artery. When using his left arm, blood had to travel *reversely* in the vertebral artery to supply the subclavian artery in the arm so that blood was 'stolen' from the brain stem, another example of vertebrobasilar ischaemia. This condition is known as 'subclavian steal' and operation to widen the subclavian artery relieves the symptoms.

Any movement that produces hyperextension of the neck, for example looking upwards, or having hair washed in hairdressing salons with the neck extended and the back of the head resting on the washbasin, is particularly prone to produce symptoms in those with severe degenerative disease of the cervical spine.

Vascular malformation

Case 15. A 21-year-old Chinese student complained of a sudden headache during the evening and went to lie down at 8.30 p.m. At 10.30 p.m. his sister went to see how he was and found that he was unconscious, having vomited. He was rushed to hospital where the casualty officer found that the patient was deeply unconscious and his temperature raised at $39°C$. His pupils were very small (pinpoint) and did not react to light. All four limbs were flaccid. The casualty officer thought that the patient's condition might have been due to a drug overdose but the patient's sister was able to exclude this possibility, and provide the important history of sudden headache, rapid loss of unconsciousness (within 2 hours), and the occurrence of vomiting. The neurological registrar was called and gave his opinion that the patient had probably suffered from a brain-stem haemorrhage. A lumbar puncture† was performed which revealed uniformly heavily blood-stained cerebrospinal fluid (CSF). The patient's condition did not improve and he died within 18 hours of the onset. At post-mortem, a brain-stem vascular malformation was found.

† See Chapter 6.

The different types of stroke

During foetal development, the primitive blood vessels develop into arteries and veins, joined by small capillaries. Failure of differentiation can leave a collection of primitive blood vessels known as a vascular malformation (see Plate 3). These are usually situated on the surface of the brain hemisphere and only rarely in the territory of the vertebrobasilar system. They can bleed in childhood or in adult life causing paralysis (hemiplegia), epilepsy, or mental defect. Although some are inoperable, the major feeding vessels to the malformations can be clipped at operation. If very large, the malformation may steal blood away from vital areas of the brain, acting as a shunt. In the case described, the rapid loss of consciousness, pin-point pupils, hyperpyrexia and flaccid bilateral hemiplegia, all suggested a brain-stem haemorrhage and the lumbar puncture helped to confirm the diagnosis.

Subarachnoid haemorrhage

Case 16. A 40-year-old labourer was drinking with his colleagues in a public house, when he suddenly complained of a severe headache at the back of the head, the worst he had ever had in his life. Whilst waiting for the ambulance, he vomited several times. On arrival at the casualty department, he was fully conscious, still complaining of the very severe headache; he was found to have marked neck stiffness and could not put his chin on his chest. Examination of the retina showed changes due to long-standing raised blood pressure. Over the next few hours he developed weakness in the left arm and leg. A lumbar puncture showed that the cerebrospinal fluid (CSF) was slightly pink in three successive samples. Further tests were performed on the CSF in the laboratory. The patient improved by the third day and an arteriogram showed an aneurysm on the right middle cerebral artery and also a smaller one on the left side.

This patient had a subarachnoid haemorrhage, which simply means blood in the cerebrospinal fluid. In 60 per cent of cases, the cause is due to a ruptured aneurysm (see Plate 4). A vascular malformation is occasionally responsible (as in Case 15) but sometimes no cause is found, possibly because the bleeding aneurysm or vascular malformation is destroyed

The different types of stroke

by the bleed or may not be demonstrated by arteriography.

Aneurysms invariably occur at bifurcations of arteries, where the muscle layer is naturally thinner. They are likened to 'blow-outs' in tyres and are more likely to occur if there is associated raised blood pressure. Although present from birth a secondary factor, such as hypertension, is necessary for them to develop. As in the case described, they can be multiple. They occur usually between the ages of 40 and 60 years but also in old age.

Typically, the onset is sudden, the headache excruciating and localized over the back of the head. The neck stiffness ('meningism') is due to irritation of the coverings of the brain (meninges) by the blood so that the diagnosis has to be distinguished from meningitis which may give a similar type of illness. (In meningitis, there is usually a warning (prodromal) illness with fever, and a lumbar puncture with laboratory tests on the cerebrospinal fluid will differentiate the diagnoses.)

An aneurysm may bleed into the brain, in which case the features are indistinguishable from cerebral haemorrhage. The patient described in Case 16, with a ruptured aneurysm, suffered from both subarachnoid and cerebral haemorrhage. The clinical signs suggested that the right middle cerebral aneurysm had bled and an electroencephalogram (EEG: see p. 54) confirmed this. The samples of cerebrospinal fluid contained break-down products of blood.

Aneurysms can be treated medically, as in this patient, or surgically. Medical treatment consists of treating the blood pressure and ensuring strict bed rest for 3 weeks. Since the mortality from subarachnoid haemorrhage is 50 per cent, and as patients are at risk from re-bleeding, surgical treatment is preferable, although not always possible, and consists of tying or clipping the neck of the aneurysm.

Subdural haematoma

Old people at home on their own are liable to falls; hitting their heads sometimes causes a blood clot (haematoma)

The different types of stroke

which is situated between the outer membrane (dura), surrounding the brain, and the substance of the brain itself. A subdural haematoma may lie quietly for many months, producing no ill-effects, but later on can give rise to headaches, drowsiness, and even hemiplegia. The original head injury may not be remembered in over half these cases.

Case 17. A 29-year-old school teacher came to the neurologist with a 3 months' history of increasing headache and weakness in his left arm. On examination, the optic discs were found to be swollen. A primary cerebral tumour was suspected and, at operation, a subdural haematoma was found. On direct questioning after the operation, the patient remembered having hit his head on a door 6 months previously. He had been momentarily dazed but quickly recovered with no ill-effects. After the operation the patient made a complete recovery.

Rare causes of stroke

Conditions which lead to increased blood viscosity can cause clotting (thrombosis) of cerebral arteries or veins. Cerebral venous thrombosis can occur in severely debilitated children, particularly with infection spreading from the ear or face. Another cause is pregnancy, where there is also an increased tendency for the development of stroke.

The pill

The contraceptive pill, particularly the high-dose oestrogen pill, has been linked with an increased incidence of cerebral venous and arterial thrombosis.

Case 18. A 24-year-old nurse was admitted to hospital whilst on holiday in Greece with headache and a feeling of general malaise. Within 48 hours, she developed obvious signs of a left hemiplegia. She had been on the contraceptive pill for 3 years. There was no family history of vascular disease. She was transferred back to this country and a CT scan confirmed the diagnosis of a right cerebral infarction. The patient made steady progress and the eventual outcome was excellent, although she still had residual weakness of the left foot. She returned to her former nursing duties after 9 months.

The different types of stroke

Polycythaemia

A blood disorder associated with an increase in the number of red blood cells, polycythaemia, can reduce cerebral blood flow and sometimes lead to arterial thrombosis. The treatment of choice is to take off a pint of blood (venesection) at frequent intervals. Leukaemia, a malignant blood disorder producing an abnormal type of white blood cell, can also lead either to brain infarction or haemorrhage.

Sickling

Sickle-cell disease is an important cause of cerebral infarction in the black population of the United States and also in Africa, although rare in this country. The red blood cells of these patients have an abnormal haemoglobin. In certain situations, particularly when oxygen is lacking, the red blood cells take on a 'sickle' appearance; these cells are then trapped in the small capillaries, blocking them, and produce infarction.

A number of different abnormal haemoglobin types have been described, which give rise to variable degrees of anaemia. Sickle-cell disease is the commonest, occurring in approximately 1 in 600 American negroes. The severe type of disease is seen in those patients whose parents are both affected with the milder type of sickle-cell disease ('sickle-cell trait'). Symptoms usually start in the latter months of the first year of life when the foetal haemoglobin (F-type) is replaced by the abnormal haemoglobin (HbS). Painful crises are due to tissue infarction in bones, joints, spleen, or abdomen. These patients are at risk from infections. Cases described in non-whites are invariably associated with combinations of HbS and other haemoglobin abnormalities. In the days before pressurized aeroplanes, sickle-cell patients were at risk from air travel.

Scurvy

Scurvy, which is caused by a deficiency of Vitamin C, was a

The different types of stroke

frequent scourge of the Navy until citrus fruits were introduced to the sailor's diet by Captain Cook in the eighteenth century. It was unfortunate that limes were chosen for the compulsory citrus fruit rations; they have, in fact, a relatively low vitamin C content, but scurvy was, nevertheless, eradicated. (This is the origin of the nickname 'limey' for the English.) Gross deficiency of Vitamin C can lead to cerebral haemorrhage.

Syphilis

Before the discovery of penicillin, syphilis frequently attacked the central nervous system, usually many years after the primary infection. Meningovascular neurosyphilis was a common cause of hemiplegia in young adults but is now rarely seen. Any cerebral artery can be affected, leading to partial or complete occlusion.

Hemiplegic migraine

A rare variety of migraine is associated with 'pins and needles' (paraesthesiae) and, occasionally, actually weakness affecting the limbs. Very rarely, cerebral infarction following a prolonged migraine attack has occurred but the weakness almost always clears within 2–3 days. Anxiety is then felt as to whether there may be a serious underlying cause such as an aneurysm or vascular malformation. Since arteriography can be hazardous, a non-invasive brain scan (see p. 55) can be helpful in this situation.

Growths (tumours)

Case 19. A 74-year-old female patient had a malignant skin tumour removed 6 years before she suddenly developed paralysis affecting her right side. Investigations were carried out to see whether she had sustained a stroke (left cerebral infarction) or whether the hemiplegia could be due to a secondary tumour in the brain, i.e. a metastasis from the skin tumour (melanoma), despite being perfectly well in the 6 years

The different types of stroke

following the operation. The CT scan (Plate 5) revealed that the hemiplegia was due to a tumour and this was confirmed at operation as a secondary melanoma deposit.

The previous medical history can sometimes be misleading.

Case 20. A 60-year-old male patient had been coughing up blood for 1 month. X-ray of the chest showed something suspicious and examination of the sputum showed malignant cells suggesting cancer of the lung (bronchial carcinoma). A partial removal of the lung was carried out followed by a course of deep X-ray treatment. 18 months later the patient had a sudden left hemiplegia, initially assumed to be due to a cerebral secondary deposit from the malignant tumour of the lung. CT scan showed that it was in fact due to a stroke and six weeks later, the patient had fully recovered from the hemiplegia.

Since secondary deposits in the brain from a malignant primary tumour elsewhere rapidly lead to death, an incorrect diagnosis would have led to the patient being considered 'terminal' but, fortunately, the patient and his family were spared this dreadful alternative. In every 100 unspecified 'stroke' cases, at least five are actually due to tumours, some of which are potentially treatable.

The great variety of the different types of stroke has been illustrated and the investigations necessary to define the size, site, and nature of the vascular damage will be described in the next chapter, before considering treatment and eventual outlook.

6

Investigations

Many people feel that 'stroke' is an acceptable diagnosis and that investigations to confirm it are not necessary. The accuracy of distinguishing the different types of stroke is low (little more than 60 per cent), but greatly improved by special tests. The previous two chapters illustrated the variations in 'stroke-like' illnesses and the problems of accurate diagnosis, which is particularly important in those cases where an incorrect diagnosis could lead to inappropriate treatment. With computerized tomographic brain scanning (p. 56), it is now possible in almost all cases to differentiate cerebral infarction from haemorrhage. What this means for the stroke patient is answered in this and the following chapters.

Two simple investigations that can be carried out to differentiate cerebral infarction from cerebral haemorrhage in over 75 per cent of cases will be described before other tests which are more 'routine'.

Lumbar puncture

This is carried out when examination of the cerebrospinal fluid (CSF) is necessary, as in patients with suspected subarachnoid haemorrhage or meningitis. A lumbar puncture is specifically contraindicated in certain situations, chiefly in patients with brain tumours with raised intracranial pressure. Lumbar puncture can be dangerous in these cases as the raised pressure in the head, together with the reduced pressure of cerebrospinal fluid can force the brain-stem downwards. For this reason, before carrying out this test, careful examination of the optic discs by ophthalmoscopy is performed to

Investigations

exclude optic disc swelling (papilloedema). The use of lumbar puncture in stroke patients is often reserved for those cases where subarachnoid haemorrhage is suspected, but there is little contraindication to lumbar puncture shortly after the acute onset of any stroke.

Patients suffering from cerebral infarction usually have clear CSF with a normal pressure. Following cerebral haemorrhage, the CSF is usually evenly blood-stained, and the pressure may be increased. This differentiation is even more accurate if further tests are carried out on the CSF, because breakdown products of blood in the CSF can be accurately detected with an instrument called a spectrophotometer, and complicated situations, for example, where bleeding may follow infarction (haemorrhagic infarction), can be more fully analysed. The accuracy of this method is claimed to exceed 90 per cent, but it will not detect bleeding into the brain substance which has not ruptured into the ventricles or subarachnoid space; in this condition, CT scanning is required. Even without spectrophotometry, CSF examination is accurate in 75 per cent of cases of stroke in separating infarction from haemorrhage.

Echo-encephalography (EchoEG)

Ultrasound is now widely used in medicine, for example in obstetrics to show the position of the baby and placenta in a pregnant woman and to monitor foetal development; the technique carries no risk to either mother or baby.

The deflection of an ultrasonic beam can be used to detect shifts of anatomical structures. An ultrasound probe, placed just above and behind the ear, can be set to register the position of the mid-line tissues, including a thin membrane (septum pellucidum), which separates the left from the right cerebral hemisphere. A tumour, for example, in the left hemisphere will push the septum pellucidum towards the right. The ultrasonic beam picks up this deflection since it has further to travel to the membrane when placed

Investigations

near the left ear, and will have less distance to travel when placed near the right ear. With this technique, the position of the septum pellucidum can be visually displayed on a cathode screen and a polaroid photograph taken of each of the measurements, first from the left and then the right side; this will quantify the degree of shift.

This method does not indicate the nature of the space-occupying lesion causing the shift. A blood clot in the brain (cerebral haemorrhage) occupies space and can be detected very quickly after it has occurred. In contrast, cerebral infarction does not immediately act as a space-occupying lesion and consequently no shift will be picked up in the first 24 hours. In this way, cerebral haemorrhage can be differentiated from cerebral infarction within the first 24 hours, with a 75 per cent degree of certainty. After 24 hours, oedema (tissue fluid) may appear around the infarcted area so that ultrasound scans performed at 48 hours and 72 hours may show successive shifts which then completely resolve by the seventh day; this means that the progress of cerebral oedema with its subsequent absorption can be monitored.

EchoEG and lumbar puncture can be easily and quickly performed after admission to hospital and, taken together, give increased diagnostic accuracy. Many hospitals may not have EchoEG facilities but, since the use of ultrasound in obstetrics is now widespread, the same basic equipment can be adapted for this purpose. There may be practical problems when two different departments require the use of the same equipment; for this reason, despite the additional cost, two machines may be preferable to one.

'Routine tests'

Not all hospitals consider all the following tests as 'routine' but examination of the blood, electrocardiography (ECG), and chest X-ray are usually considered to be mandatory when any stroke patient is admitted to hospital.

Investigations

Blood tests

Examination of the blood will detect blood conditions, e.g. anaemia or polycythaemia (see p. 47) both of which, because they affect cerebral blood flow, can produce a stroke. Other blood conditions, such as leukaemia, sickle-cell disease, and vitamin deficiencies will also be detected in the blood. Blood glucose estimations are necessary to detect the presence of diabetes. Enzymes in serum (which is the fluid of the blood after the cells are removed) act as indicators of damage to heart muscle or lung and can be raised following stroke. One such enzyme, creatine phosphokinase (CPK), produced by muscle and brain, is found in increased concentrations in the serum in muscle diseases but also in the CSF of patients following cerebral infarction, and may be correlated with the amount of brain damaged.

Electrocardiogram (ECG)

The ECG machine is used to measure the electrical activity of the heart and the paper trace provides a permanent record. It should always be done in stroke patients in order to detect underlying heart disease, and to see whether the patient may have had a recent heart attack (see Case 7, p. 36). Disordered rhythms of the heart are also detected by the ECG; they are not uncommon and may require treatment. Patients recovering from a stroke are susceptible to heart disease in later years, so that this is another reason to perform an ECG. Following subarachnoid haemorrhage, the levels of certain chemicals circulating in the blood are increased, e.g. catecholamines (noradrenalin and adrenalin) and these can give rise to very abnormal ECGs; during the recovery phase, the ECG usually returns to normal, providing there is no underlying heart disease. In patients with long-standing high blood pressure (hypertension) the ECG will also show electrical evidence of enlargement of the left ventricle.

Investigations

Chest X-ray

This, too, will show an increase in heart size in hypertensive patients. It is also important if a cerebral tumour is suspected, since it may show a primary cancer of the lung.

Skull X-ray

A plain X-ray of the skull occasionally provides useful information in stroke patients, since brain tumours are sometimes calcified. It may also reveal a shift of the pineal gland, which is a small gland of uncertain function, situated in the lower part of the brain, that calcifies in 50 per cent of normal people by the age of 50 years. A shift of this on X-ray indicates a space-occupying lesion, e.g. cerebral haemorrhage, although the EchoEG is much more accurate in this regard. Another use for skull X-rays in stroke is to detect fractures, e.g. of the facial bones. This possibility should be considered in any stroke patient who has facial bruising or with a history of a fall following the stroke. Calcification in the internal carotid artery may also be seen on the skull X-ray; this calcified atheromatous material is an indicator of arteriosclerosis.

Electroencephalography (EEG)

The normal electrical activity of the brain can be shown by means of the EEG. Its main use is to detect electrical disturbances, such as those that occur in patients with epilepsy, and to detect electrical differences between the two cerebral hemispheres. The EEG can rarely tell the exact nature of the trouble causing the electrical disturbance, for example it cannot with certainty tell the difference between cerebral infarction and a cerebral tumour. In patients with multiple intracerebral aneurysms (see Case 16, p. 44) it can be used to detect which aneurysm has probably bled.

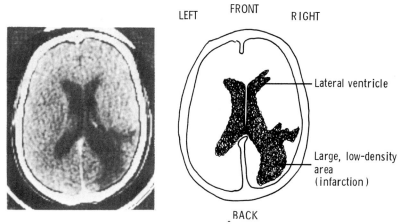

PLATE 1. CT scan of Case 8 (p. 37) showing a low-density area towards the back of the right cerebral hemisphere due to infarction of brain tissue.

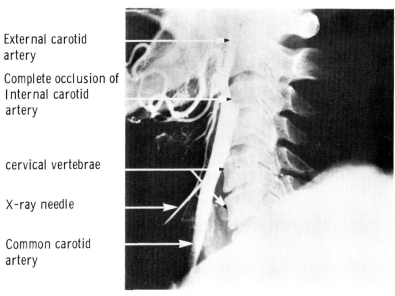

PLATE 2. Left common carotid artery angiogram of Case 10 (p. 40). An X-ray needle is introduced into the common carotid artery, and X-ray contrast medium injected. This patient has a complete blockage (occlusion) of the left internal carotid artery.

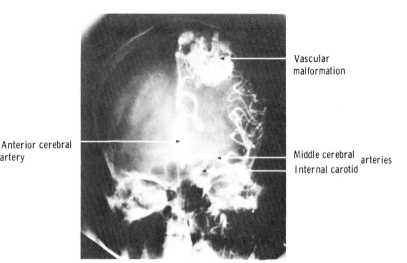

PLATE 3. Vascular malformation (p. 43). A left carotid artery angiogram has been performed which demonstrates a large malformation supplied chiefly by the anterior cerebral artery, which is dilated.

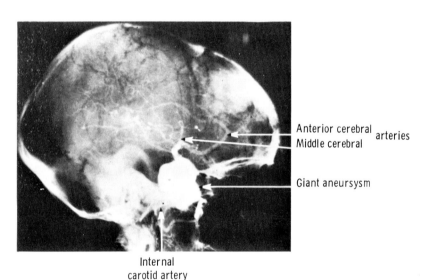

PLATE 4. Intracranial aneurysm (see p. 44). A giant intracranial aneurysm is demonstrated by a right internal carotid artery angiogram. The aneurysm arises from the internal carotid artery and is pressing on the back of the eye. The anterior and middle cerebral arteries are both beginning to fill. The patient was a 70-year-old woman with progressive eye symptoms.

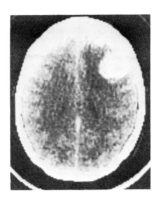

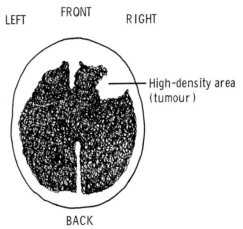

PLATE 5. CT scan of Case 19 (p. 48) showing a high-density area in the front of the right cerebral hemisphere produced by a brain tumour.

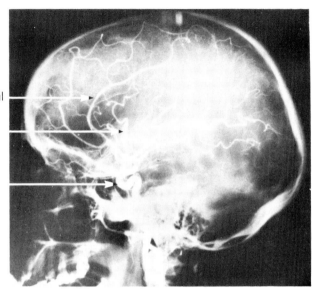

(a)

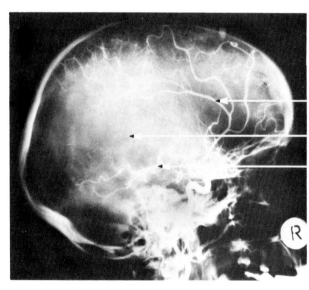

(b)

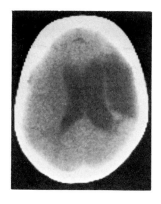

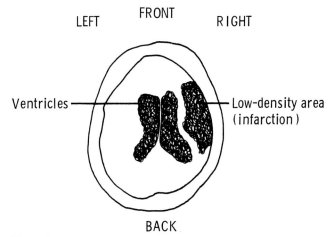

(c)

PLATE 6. The patient was a 44-year-old woman who had sustained a left hemiplegia. Angiography ((a) and (b)) and CT scan (c) were performed. Angiography shows which artery is blocked, and the CT scan shows the area of brain damaged.

(a) Left internal carotid artery angiogram. The left internal carotid artery has been selectively catheterized (see p. 57) and X-ray contrast medium injected to show the intracranial arteries supplying the left cerebral hemisphere. (b) Right internal carotid artery angiogram. This time, occlusion of the middle cerebral artery was demonstrated and no filling of the branching arteries in this territory was seen. (c) CT scan shows a low-density area in the right cerebral hemisphere due to infarction of brain tissue. This area of brain is supplied by the middle cerebral artery.

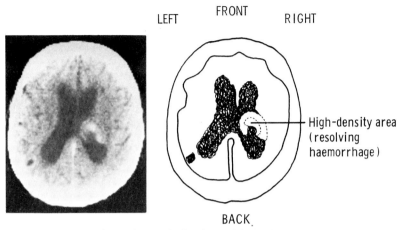

PLATE 7. CT scan of Case 23 (p. 72) showing a high-density area deep in the right cerebral hemisphere due to haemorrhage into the brain tissue.

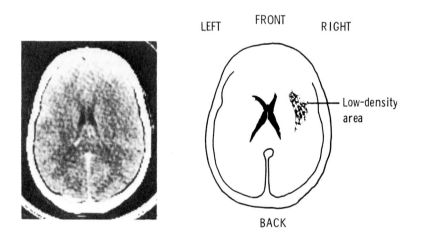

PLATE 8(a)

(b)

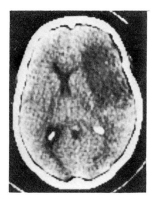

LEFT   FRONT   RIGHT

Ventricles compressed, pushed to the left

Large low-density area

Ventricles (posterior part)

BACK

PLATE 8. Serial CT scans of Case 25 showing development of cerebral oedema and resolution after treatment with dexamethasone (p. 75).

(a) CT scan at 18 hours. Low-density area in right cerebral hemisphere due to infarction of brain tissue. (b) CT scan on day 6. Large low-density area due to brain-swelling (cerebral oedema) with compression of the ventricles, particularly on the right side, and shift to left. (c) (*next page*) CT scan on day 14. Residual, low-density areas with resolution of the brain-swelling. Ventricles are now in the normal position.

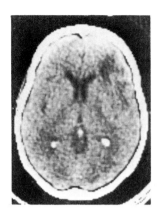
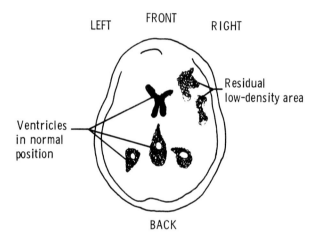

PLATE 8(c)

Investigations

Brain scan

Before a brain scan can be done, a synthetic radio-isotope (technetium-99 m-pertechnate) is injected into a vein, and a special (gamma) camera is used over the skull to show where the isotope is taken up. (Because the thyroid gland would normally take up this substance, its function is temporarily 'blocked' by giving a tablet by mouth (oral potassium iodide)).

The normal brain does not take up any isotope and appears on the scan as a blue (cold) area. Because of their high concentration of blood vessels, face and scalp areas show up as red (a 'hot' area). If the barrier between blood and the brain is disturbed by a stroke or tumour, there will be a large area of extracellular fluid space (oedema) in these regions where the density is greater than the surrounding brain. As a result, more isotope is taken up in these areas than the surrounding brain, giving a positive scan, which is shown by a yellow to red 'hot' area within the brain substance.

If a brain scan is carried out immediately after a stroke due to cerebral infarction, it is usually negative, although up to 40 per cent of scans may be positive within the first 48 hours. Because of oedema, the brain scan becomes positive within the first 7 days, staying positive for several weeks and clearing by 3 months. In contrast, a cerebral tumour will be shown on the first scan and will become increasingly positive. In this way, although the isotope brain scan cannot differentiate cerebral infarction from haemorrhage, the time-course of events and use of repeat scans can differentiate cerebral infarction from cerebral tumour, the course of the illness being quite different in the two diseases. A cerebral haemorrhage will also be demonstrated on the brain scan but the appearances are identical to that produced by infarction. Subdural haematomas (p. 45) can be identified with a high degree of certainty by the isotope brain scan.

If the rate of clearance of the isotope is used as an additional

Investigations

measure (called a 'dynamic' brain scan as opposed to the previously described 'static' scan), this method can then be used to detect a vascular malformation. The 'hot' spot is clearly shown and the rapid shunting of blood through the malformation will be indicated by subsequent scans. A comparison of the clearance rates of isotope by the two hemispheres can also be used as a measure of narrowing (stenosis) in the extracranial (carotid) arteries since, when this is present on one side, there will be delay in the cerebral blood flow in the hemisphere of the same side, providing there is no significant disease in the carotid artery on the opposite side.

Computerized axial tomography (CAT or CT scanning)

The application of CT scanning to medicine has been described as the greatest advance in medicine since the discovery of X-rays. The two principles involved in CT scanning had separately been known for a long time, but it was Godfrey Hounsfield, an engineer, who in the 1960s first thought of combining them. Working for EMI, he produced the first CT scanner, so that the initial scans were known as EMI scans. With the advent of different models of CT scanner produced by other companies, the term CT scan was introduced. Because of controversy regarding their usefulness, it is worth considering the medical and economical applications of CT scanners and their relevance to stroke patients.

Tomography is used frequently in radiology. When a chest X-ray is carried out, the dose of radiation and the position of the X-ray plate is fixed to provide a standard picture or 'plain chest X-ray'. By altering the X-ray beam, the penetration and hence the focusing can be altered in the same way as a camera lens. A series of pictures can then be taken at different depths of focus, thereby producing a series of sectional pictures (like a sliced loaf). This is the principle of tomography, which is frequently used to show more clearly, for example, a lung cancer (carcinoma of the bronchus).

Investigations

The second principle involved in CT scanning is that different tissues absorb electrons or X-rays to different degrees, i.e. each type of tissue has a different absorption coefficient. Because of this, tumour tissue, haemorrhage, infarction, CSF, abscess, and bone, can be differentiated since each has a different absorption coefficient. The technique is so accurate that grey brain matter can be differentiated from white brain matter, and anatomical shifts are clearly shown, as is the presence of cerebral oedema. The great advance of CT scanning was that not only could the position of the lesion be shown but also its probable pathological nature could be demonstrated. CT head scanning involves taking serial sections (tomography) through the brain and measuring the absorption coefficients of the different tissues in each 'slice'. The computer presents the information on a cathode screen as a cross-section of the head, showing skull and brain, and the pictures are developed to provide a permanent record. A number of standard serial cuts are taken and, by injecting an X-ray medium into the blood, enhancement of the brain lesion may help to render it more obvious. Examples of CT scans are shown in the Plate section of this book.

An exciting new development is positron emission tomography, where the beam consists of positrons instead of electrons. There is less scatter of the positron beam and even higher resolution will be possible with this new equipment in the years to come.

Angiography

The definitive technique for demonstrating blood vessels is by performing a special X-ray called an angiogram (or arteriogram) but this procedure is not without risk, having a morbidity of over 1 per cent, even in experienced hands. Timing of the angiogram may be crucial and it should only be done if specific treatment depends on its results, e.g. carotid endarterectomy (see page 102).

The procedure involves introducing either a needle directly

Investigations

into one of the vessels in the neck and injecting a special X-ray-opaque dye, or by introducing a catheter via the artery in the groin (femoral artery) and manoeuvring it to the origins of the vessels in the neck under X-ray viewing control. This last method removes the need for separate needle punctures of each of the arteries in the neck and all the arteries to the brain can be visualized on X-ray plates. After injection of dye into the catheter, a series of automatic X-rays are taken at 1-second intervals (see Plates 6 and 7). This demonstrates the arterial or early-filling phase and the late-filling or venous phase. The procedure often involves a general anaesthetic with its attendant risks. Unfortunately, CT scanning has not obviated the need for angiography in selected patients.

Non-invasive angiography

New techniques have been developed using both ultrasound and the Doppler effect for detecting the direction of flow in arteries. These techniques (e.g. carotid phonoangiography and oculoplethysmography) have been introduced as a method for measuring and following up patients with arterial disease. The great advantage is that these techniques are 'non-invasive' and without risk, but their disadvantage is that they do not show narrowing of arteries which are only slightly stenosed (up to 40 per cent); above 50 per cent of narrowing, the techniques are very good even compared to angiography. Further developments will produce better 'non-invasive' methods but angiography is still the investigation of choice for symptomatic disease of the neck arteries.

7

Who looks after the patient in hospital?

There is a sharp contrast between the treatment of the *suspected* stroke patient and the *suspected* heart attack patient. For the latter, the klaxon of the emergency cardiac ambulance signals the importance of speedy transfer from home, office, or street to the nearest coronary care unit. In contrast, 'tomorrow will do' has often been the axiom for the stroke patient.

There are many reasons for this difference in attitude. For example, heart attacks affect younger people and, until recently, treatment was always considered to be better in hospital than home. Coronary care units took pride of place in the new general hospitals of the 1960s and 1970s because, in these units, disorders of heart rhythm could be easily monitored and appropriate treatment instigated.

Because it has yet to be shown whether it is possible to limit brain damage immediately following a stroke, stroke patients are not given the same chance as, or the same level of care as, the patient who has had a heart attack. There are remarkably few stroke units but, in a recent report on three, it was found that the risk of developing complications secondary to the stroke illness, such as bronchopneumonia and pulmonary embolism, was very significantly reduced in those patients admitted to the stroke unit compared with patients admitted to a general ward. Since these complications are potentially avoidable, this implies that a higher standard of care is given to those patients admitted to the stroke unit.

In the last decade the need for more specialized care has been recognized and developed in the hospital and in the community. No longer is it just the doctor and nurse who look after the patient. The hospital team is made up of the doctor, the nurse, the physiotherapist, the occupational

Who looks after the patient in hospital?

therapist, the speech therapist, the social worker, and the psychologist. Each has an important part to play in the management of the stroke patient and their individual roles will be discussed in this chapter. The work of the psychologist will be discussed in Chapter 10).

The doctors

There are a number of different medical specialities involved in the care of different types of diseases, for example nervous diseases (neurology), heart diseases (cardiology), respiratory diseases (respirology), kidney diseases (nephrology), gastrointestinal diseases (gastroenterology), joint diseases (rheumatology). Geriatrics, or the study of diseases of the elderly, is a more recent speciality. Rehabilitation is not yet fully accepted as a specific speciality in this country although training programmes are operational in several European countries and North America. In district general hospitals, many general physicians cover all these specialities, referring patients to specialists when this is considered necessary. Most teaching-hospital consultants are specialists in one of the disciplines listed above, but which specialist is interested in stroke? The acute stroke patient is usually admitted under the care of a general physician or a geriatrician, whilst the recovering stroke patient might be referred to a rheumatologist with an interest in rehabilitation. As there is no 'strokeologist', the acute stroke patient is not considered to be within the interest of any one specialist, although the neurologist might be consulted if the diagnosis is in doubt. This is clearly a barrier to the formulation of suitable guidelines to the investigation, treatment, and follow-up of stroke patients and may account for the lack of progress in the management of stroke. There are hopeful signs that changes are occurring: in certain hospitals, the neurologists, neurosurgeons, and vascular surgeons, joined by the paramedical departments, combine as a team, as it is appreciated that no single person can fulfill all the requirements needed to look

Who looks after the patient in hospital?

after the stroke patient, and a multidisciplinary approach is essential. The role of these different personnel is expanded in later sections.

The nursing staff

The devotion of nursing staff towards patients in their care is well-recognized. Not all patients are courteous, yet their nurses will still care for them as well as for any other in the ward. Antisocial behaviour can occur because of organic brain damage, but it can also be a demonstration by the patient that he is unhappy with his lot, because he may feel that he is not progressing satisfactorily, and, after weeks of intensive efforts aimed at helping him regain his independence, the patient may become demoralized and start to regress.

Most patients fully appreciate the efforts made by the nursing staff and try to help themselves as much as possible; they are rarely demanding and more often will call a nurse because another patient is in difficulties.

Staff shortages often put an unbearable pressure on the nurses who are only too well aware of the times they cannot provide of their best. For example severely disabled patients requiring two-hourly turns to avoid bed sores may have to be left for longer intervals between turns. Bed sores can take a long time to heal in a disabled, debilitated patient and will delay rehabilitation and eventual discharge.

Rehabilitation of the stroke patient should begin as soon as possible after the onset of the acute illness but nurses may not have as much time as they would wish to spend helping disabled patients to perform simple everyday tasks. It may be quicker to feed a patient than actually help him learn to feed himself again, particularly if there are six other patients on the ward waiting as well. In some European and Asian countries, many of the everyday nursing tasks are carried out by relatives coming in daily, spending several hours with the patient.

Who looks after the patient in hospital?

The ward sister or her deputy puts aside definite times during the week to discuss their patient's progress with his relatives. Depending upon the patient's clinical condition and rate of progress these talks may be more frequent at certain times than others, when practical information, explanations concerning treatment, and answers to relatives' queries can all be given. Relatives often confirm that what the nurse has told them was already said by the house physician (resident) at an earlier meeting, an indication that the doctors and nurses work as a team and, as a team, try to provide the same accurate information.

The physiotherapist

The number of physiotherapists attached to a hospital varies enormously. They treat patients in the ward and also in their own physiotherapy department. In addition to their in-patient work-load, they have a large out-patient commitment and have to treat a large number of patients with many different complaints. They do not automatically see every disabled patient in the hospital. Each has to be referred initially by the consultant in charge of the case, and the timing of this referral may be crucial.

In the early days after a stroke, the physiotherapist will usually provide chest physiotherapy to prevent chest infection and will apply gentle, passive movements to paralysed limbs. The amount of treatment-time that each patient receives on the ward is very limited because the physiotherapist often has a large number of ward patients to see, surgical as well as medical. Once the patient begins to be mobile, two physiotherapists are needed to help with the first few walking steps; they instruct the nurses how to 'walk' the patient and, after the treatment period by the bedside, the nurses try and continue the physiotherapist's regime during the day.

Once the patient is well enough to go to the physiotherapy department, the length of treatment can be increased, depending upon the patient's physical tolerance. The physiotherapist

Who looks after the patient in hospital?

may then work with a group of similarly disabled patients, and this group therapy, because it means leaving the ward and meeting other patients at various stages of their recovery, is a great morale-booster for the patient. After their first physiotherapy sessions in the department, patients return to the ward often feeling very tired, unwilling to do much for themselves for the rest of the day. This low physical tolerance gradually improves, but is clearly related to the age of the patient and the physical ability prior to the stroke.

Physiotherapy treatment can be continued on an outpatient basis but its value must be carefully weighed against the problems of transport and whether continued treatment is really necessary. During a period of levelling out, when nothing seems to be happening, it is difficult for the stroke patient to accept that extra physiotherapy will not speed up the rate of recovery.

The occupational therapist

Old ideas die hard: the traditional image of the occupational therapist as someone who instructs patients in basket-weaving is a long time passing. Patients referred to the occupational therapist are assessed initially to see what they can do, before a treatment programme is devised for that patient's own particular requirements. Stroke patients may be referred to the occupational therapist two weeks after admission. Occupational therapy treatment is likely to be limited at this point of recovery in the stroke illness but it is important to involve the occupational therapist in the patient's treatment programme as soon as possible.

Once a patient has started to put his weight on his affected leg, the occupational therapist will help him to transfer from his bed to the chair and back. She will advise and teach the patient how to dress himself and cope in his toilet and washing. She is an expert in the use of aids for the patient in hospital or at home. Special cutlery, non-slip mats and special cups can be of considerable help while there are

Who looks after the patient in hospital?

feeding difficulties. Hoists can help when transferring the patient from bed to chair and also into the bath.

The programme for each patient has to be specifically devised to meet the patient's requirements. The first priority is to make the patient independent in his activities of daily living (feeding, dressing, washing, toilet), and to ensure that his walking is safe. It may not be possible to proceed further than this, yet this level of function will be enough to allow an elderly patient, living alone, to manage at home with the aid of social services. It is essential for a working man to proceed much further along the rehabilitation road if he is ever to regain his employment. As soon as the patient is well enough to go to the occupational therapy department, he will be able to take part in more precise training programmes.

Activities to promote hand grip and dexterity can be learnt in the light workshop. Later on, some male patients transfer to a 'heavy' workshop and this will entail activities which will help to build up the power in the affected limbs. Facilities vary considerably from hospital to hospital, as do rehabilitation centres. The graduated exercises, involving safe, workshop equipment with different degrees of difficulty and under careful supervision, help the patient to regain his confidence and encourage him to persevere towards his goal of readjustment after a stroke.

Case 22. A 60-year-old taxi-driver was admitted to hospital after a stroke had caused weakness, but not total paralysis, of his left arm and leg. After 4 weeks, he was transferred to a rehabilitation centre, and at this time he could walk without a stick, but still had difficulty with hand movements and weakness of the shoulder. After 3 weeks at the centre, he was discharged home and asked to attend three times a week for out-patient physiotherapy and occupational therapy. His only remaining problem was the shoulder weakness and fine movements of his left hand. He was completely independent in self-care, and could travel on public transport.

Three months after his stroke, he tried driving his taxi and had no problem, but he could not manage one activity essential to his trade: he could stretch out his left arm but he could not flick down the taxi's

Who looks after the patient in hospital?

clock-meter which he would have to do when he was operational again. It was almost 6 months after the stroke before he was able to do this and only then could he go back to work.

Occupational therapy and physiotherapy are the two main aspects of rehabilitation and both types of therapists work closely together. Both are looking at functional recovery in different ways and their types of assessment are quite different. The physiotherapist looks at movement in terms of muscle-power, muscle-tone, muscle-bulk, and coordination. The occupational therapist looks at the ability to perform everyday tasks essential for maintaining an independent existence and, if possible, will visit the patient's own home to see if any adaptations or alterations are necessary before discharge. She will liaise with her colleague, the area occupational therapist, who works in the community, and with social service departments to provide the necessary help in order to carry out any alterations or supply aids.

The speech therapist

About one-third of all stroke patients may have their speech affected following a stroke, and in one of two main ways. Either the content of speech is affected so that patients may not be able to formulate words or sentences or understand what is said to them (aphasia), or the content of speech is normal but the words are slurred owing to weakness of the face or tongue (dysarthria).

Speech therapists undergo a 3-year training programme and some schools now are working towards a degree course. Their work is in three main areas: stroke patients, patients who have undergone laryngeal or throat surgery, and children with speech problems. There is a major shortage of speech therapists in the U.K. and many district hospitals have the services of only one speech therapist or even 'half a therapist', i.e. 2 or 3 days a week only, whilst for the rest of the week she is based at another hospital or children's unit.

Who looks after the patient in hospital?

Not all the doctors appreciate the value of referring patients to the speech therapist, who must first assess the patient and diagnose the nature of the speech problem. Patients with dysarthria usually improve quite quickly but if there are also swallowing problems, then the task of re-education can be very difficult and very time-consuming. Patients with different types of aphasia (Chapter 3) have different chances of recovery. Treatment can be on an individual basis or in a group and is first directed at communication, i.e. enabling the patient to make himself understood and then to capitalize on the spontaneous recovery of language function. Relatives need to know how best to help the patient at home and the formation of a speech club, specifically for relatives, provides an excellent forum for this purpose (p. 123).

The social worker

Social workers in the U.K. are authorized by the local area health authority and are based either in the community or in the hospital where they are attached to one or more medical firms. They see those patients referred to them by the consultant but patients or relatives may request to see them directly because of financial problems, collection of pensions, benefits, etc.

The role of hospital social workers has changed considerably since the days of the 'medical almoner'. Being part of the medical team, they are aware of the medical as well as of the social problems of the patient. At case conferences or on ward rounds, they learn of the patient's disability and his progress. Together with the occupational therapist, the social worker can make a home visit, prior to the patient's discharge, and arrange community-based domiciliary care as required, e.g. meals-on-wheels, home help.

Once the patient is home again, the social worker will maintain contact, ensuring that he and his family are receiving services and managing adequately. The social worker can also

Who looks after the patient in hospital?

instigate re-referral to the hospital doctor, or inform the general practitioner if she feels that the patient's medical condition has deteriorated.

A common problem is how to cater for patients, usually living alone before their stroke, who are now severely disabled and have no chance of returning to an independent existence. The social worker devotes considerable time to the small group of stroke patients who fall into this category. In the hospital situation, this group is expanded considerably by patients with a wide variety of medical conditions, as well as patients who are simply aged and can no longer manage at home on their own. An increasing number of hospital beds are filled with such patients, and until the social worker can find the appropriate, alternative placement, for example warden-controlled flats or homes or long-term institutional care, the patient is deemed to 'block' a hospital bed.

A patient can be discharged too soon or may be kept in hospital too long and both can be equally disastrous. Multi-disciplinary teams looking after stroke patients understand each other's problems and their unified approach works for the patient's benefit.

8

Admission to hospital

There is no standard medical treatment for stroke nor is there even agreement about whether a patient is best looked after at home or in hospital. Because stroke patients may stay in hospital a long time, there is a feeling that little can be done in the way of treatment and this has led to a general air of pessimism. The general practitioner faced with a patient who has had a stroke often wonders what to do—the 'doctor's dilemma'. His management will be governed not only by his training and his own experience, but also by the ancillary facilities available. His first decision, whether to treat the patient at home or in hospital, is not determined by these factors, but usually by whether the patient or his family would prefer hospital admission or not. Although this puts the initial decision-making in the hands of the family, it does not resolve the 'doctor's dilemma', because this, in turn, can create problems in the subsequent days and weeks. Sometimes the doctor has no choice but hospital, but this is irrespective of the condition of the patient, since there is little evidence that more severely affected patients are admitted to hospital. Hospital admission is often determined by social factors, for example whether the patient is living alone or whether social services are available. The typical, wait-and-see approach of the family and the general practitioner is highlighted by the following case history.

Case 21. A 64-year-old business man had gone to bed early because he had felt unwell that evening. During the night, he went to the lavatory and found that his right leg was heavy and that he could not walk properly. He woke his wife who helped him back into bed. At 7.30 a.m. the next morning his wife tried to rouse him but found that he could not speak and was only moving his left arm. Before telephoning the general practitioner's surgery, she waited until 9.30 a.m., firstly

Admission to hospital

to see whether her husband might improve, and secondly to ask the doctor to come and see him and confirm her own view that her husband had had a stroke. The general practitioner visited the patient at 11.15 a.m. and found him unable to speak and unable to move his right arm or leg. In a separate room, the doctor asked the wife whether she wanted to try and manage her husband at home or whether she favoured hospital admission. The wife's answer was to put the same question to the doctor, who suggested that she should look after her husband at home for a few days to see how he progressed, and then re-assess the situation. After three days, the wife could no longer cope, and the patient was admitted to hospital.

In this case, which is very typical, the patient could not speak for himself. His wife might have categorically said at the beginning that she could not have managed her husband at home under any circumstances, but in fact the doctor answered his own question according to his experience of similar patients. There was no suggestion that the wife should have telephoned the doctor any earlier, the inference being that this would not have made any difference. The attitude of both the patient's wife and the doctor was 'let us see what happens', the assumption being that nothing could be done for the acute illness. The wife was initially unaware just how much nursing care her husband would need. How would she manage feeding, toileting? How could she leave the house? Who would help her with groceries, etc? In view of these factors, it is not surprising that most stroke patients are admitted to hospital *after* 24 hours, and usually between 48 hours and 1 week, simply because the family cannot manage. This delay in admission has important consequences in the initial treatment of stroke (p. 74).

In hospital, different medical 'firms' or teams take it in turns for looking after medical emergencies admitted. A medical team is usually responsible for a 24-hour period, e.g. 9 a.m. to 9 a.m., when another team takes over. The number and type of medical admissions have varied from decade to decade, owing to the prevalence of the various diseases in the community and the changes that have occurred because of improved housing, working conditions, and

Admission to hospital

education, as much as advances in new medical treatment. In a typical 'acute medical take' of recent years, there would be two or three patients with either proven or suspected heart attacks (coronary thrombosis), one patient with heart failure due to pre-existing heart disease, two or more patients (depending upon district and day of the week) who had taken an overdose, i.e. attempted suicide, one or two patients with chest infection, a patient in diabetic coma, one or two stroke patients, and one elderly patient found neglected at home, and admitted for social reasons.

The house physician who first admits the patient obtains the history and does a physical examination, then initiates investigations and treatment according to accepted medical priorities. The interest in intensive-care monitoring of patients with acute coronary thrombosis in recent years would put these patients high on the doctor's list of priorities. The definite guidelines for the emergency treatment of patients with respiratory problems or patients who had taken an overdose, or in patients in diabetic coma, would automatically mean that they would be treated as urgent cases. Stroke patients as well as the patient with 'social problems' would be considered as 'low priority'. Because definite guidelines for their management have not been laid down, there is a lack of urgency in the initial treatment of stroke patients.

These emergency medical cases are admitted either from the casualty department, or from the general practitioners in the district after discussing the case with either the admitting house physician or his immediate superior, the medical registrar. The selection of emergency cases for admissions is made against a constant background problem of shortage of beds, either due to ward closures from lack of finance or staff holidays; other reasons are that medical wards have too many beds 'blocked' by long-stay patients who cannot be discharged because of social problems, the pressure from waiting lists, and the need to admit patients urgently from out-patient clinics for investigation. These difficulties usually mean that doctors may be reluctant to admit stroke patients

Admission to hospital

to hospital, an attitude reinforced by the fact that there may already be three or four stroke patients in a medical ward of 20–30 patients. The general practitioner is well aware of the hospital doctor's dilemma and is often apologetic at having to refer a stroke patient. When on the ward round, the consultant, the 'chief' of the medical firm, sees yet another stroke patient admitted, his major concern is how to get his 'blocked beds' moving and these patients home.

Treatment on admission to hospital

It may be necessary to carry out first-aid measures, such as 'protecting the airway' before the doctor on duty at the hospital has time to interview the ambulance team and accompanying relatives.

History

Providing the stroke patient is not unconscious or has difficulty with speech due to dysphasia (p. 18), an accurate history from the patient is essential and will either confirm the diagnosis of 'presumed stroke' or raise other diagnostic possibilities, such as a brain tumour or subdural haematoma. The past medical history, family medical history, and social history are also taken. If the patient is brought to hospital alone, then the circumstances in which the patient was found is obtained from the ambulance team, and this is also of medico-legal importance. An accompanying relative must always be interviewed and, ideally, should accompany the patient to the ward before returning home.

Physical examination

A careful examination by the doctor will confirm the presence of a *focal neurological deficit*, i.e. damage to one area of the brain causing loss of function in the corresponding part of the nervous system. The presence of dysphasia and a right

Admission to hospital

hemiplegia will confirm that the *site of the lesion* (that is, which part of the brain has been affected) is in the left cerebral hemisphere. But is the lesion due to a vascular insult (infarction or haemorrhage) or is it due to something else? Careful examination of the other body systems is essential, in particular, whether the patient has hypertension, or heart disease. Pulse, temperature, and blood pressure are routinely taken by the nurses in casualty and serve as baseline readings.

Investigations

These have been described in Chapter 6 and the reasons for performing them discussed. Initial investigations include electrocardiogram, chest X-ray, and blood tests but the results will take hours or even days to emerge from the laboratory, depending upon the tests requested. Neurological investigations within the first 24 hours include lumbar puncture to examine the cerebrospinal fluid and, if possible, echoencephalogram and CT scan. In this way it should be possible to determine the *site of the lesion* and the nature of the lesion. Hopefully a definite diagnosis can then be made as illustrated in the following, straightforward case.

Case 23. A 67-year-old man was admitted to the casualty department 6 hours after the sudden onset of weakness affecting his left arm and left leg. He had become increasingly drowsy and confused so that the history was obtained from his wife. Examination of the patient revealed a left facial weakness and left homonymous hemianopia (page 21). Examination of the other body systems was normal apart from a slightly raised blood pressure. There was no history of known hypertension or evidence from examination of the eyes or heart that he had suffered from long-standing raised blood pressure. The chest X-ray and the electrocardiogram did not show cardiac enlargement, confirming this opinion. It was considered safe to perform a lumbar puncture and the cerebrospinal fluid was evenly blood-stained. An echoencephalogram showed a shift of mid-line structures to the left suggesting the presence of a right cerebral hemisphere mass lesion. A CT scan was performed (Plate 9) which demonstrated a deep right cerebral hemisphere haemorrhage. The diagnosis was made of a 'right cerebral hemisphere haemorrhage causing a left hemiparesis'.

9

The first few days

Monitoring progress

One of the most important duties for the medical and nursing team is to determine whether treatment is actually working. This means that the patient requires repeated assessments at regular intervals of time to monitor progress. 'Quarter hourly observations' can be performed on a general ward by the nursing staff or, in an intensive care unit, continuous monitoring is possible. If necessary, this can be supplemented by instrumentation—chest leads attached to the patient for monitoring the heart beat or respiratory pattern and even equipment to measure the blood pressure, second by second. Since this type of monitoring is extremely time-consuming for the nursing staff, the medical team must have their objectives clear.

The most important parameter to assess is the patient's level of consciousness since this relates to immediate prognosis (Chapter 15). A deepening level of consciousness can be due to an extension of the area of haemorrhage or infarction but often it is due to the presence of brain swelling (cerebral oedema) around the area of damaged brain. The patient with increasing brain swelling will become drowsy, then unresponsive to commands and finally to painful stimuli. Secondary effects of the brain swelling will also be seen so that the patient's pulse may become irregular, his breathing pattern alter and later his blood pressure rise. With correction of the brain swelling, these changes can be reversed.

Case 24. A 54-year-old man was admitted following the sudden onset of weakness affecting the right arm and leg. Investigations confirmed the presence of a left cerebral haemorrhage. He was drowsy on admission but 3 days later his level of consciousness worsened in that

The first few days

he became unconscious and not reacting to painful stimuli. His pulse rate was raised but now occasional irregular (ectopic) heart beats were noted. His respiratory rate and pattern were irregular and his blood pressure began to rise. One pupil became enlarged and unreacting. The patient had been treated with dexamethasone (p. 75) which may have delayed the development of brain swelling but clearly it was no longer effective in the prescribed dosage. An intravenous infusion of mannitol was given over half an hour. Within a short time, the blood pressure began to return to normal as did the respiratory rate and pattern. After fifteen minutes the heart beat stabilised and after 20 minutes the dilated pupil became responsive. The patient's level of consciousness improved over the following one hour.

This case illustrates the importance of monitoring vital functions, in particular, the level of consciousness.

Drug treatment

Following cerebral infarction and haemorrhage, brain swelling commonly occurs. The leakage of blood into the brain in cerebral haemorrhage quickly acts by its mass effect to compress adjacent brain areas. The amount of brain swelling that results is variable and may not be seen with small intra-cerebral haematomas. In contrast, even small areas of brain infarction may, on CT scanning, be seen to be surrounded by an extremely large area of brain swelling. There is still some controversy as to the frequency with which this brain swelling or oedema occurs. Those patients dying within days following a cerebral infarction appear to have considerable brain oedema, which, by exerting pressure on the brain, leads to a deepening level of consciousness and can subsequently cause brain-stem compression, an effect that can be seen at autopsy. Since we believe that immediate mortality following cerebral infarction is related to the development of brain oedema, physicians interested in stroke have searched for many years for a drug or treatment regime which might reverse brain oedema. The ideal treatment would be one that could be started from the outset of the stroke with the aim of limiting the area of brain damage and *preventing* the development of brain oedema.

The first few days

Dexamethasone

Steroids are powerful drugs, used widely in many branches of medicine because of their anti-inflammatory effects. How steroids reduce brain swelling is not known but they are thought to stabilize cell membranes preventing further fluid build-up within and around cells. A number of steroids have been tried in the treatment of stroke patients, but dexamethasone is the most widely used, and can be given orally or by injection into a vein or muscle. Dexamethasone is known to be very successful in the treatment of brain swelling seen in association with brain tumours and following head injury. Since, in some patients with brain tumours, clinical benefit appears to occur before there is any dramatic change in the CT scan, this suggests that it may have other actions, possibly directly against tumour cells. As already mentioned, the results of trials using dexamethasone in acute cerebral infarction and haemorrhage have been inconclusive. It is possible that dexamethasone is *only* successful in those patients shown to have brain oedema.

Case 25. A 58-year-old woman was getting off the 'bus at 8.40 a.m., on the way to work, when she suddenly felt numbness in her left leg, and both legs gave way. A fellow passenger caught her before she fell to the floor of the 'bus. She realized that she could not speak properly. An ambulance was called and, in the casualty department, she was found to be drowsy, with a flaccid left hemiplegia, and slurred speech due to left facial weakness. A bruit was noted over the left subclavian artery.

A lumbar puncture was performed but the cerebrospinal fluid was normal. A CT scan (Plate 8(a)) was performed within 24 hours which showed an area of diminished density, but no oedema. A diagnosis of a 'right cerebral hemisphere infarction causing a left hemiplegia' was made. The patient became even more drowsy, asleep most of the time, but reacting to commands.

Dexamethasone was started on the sixth day of admission. By the following day the patient was alert, talking, and much better, although the weakness in the left arm and leg had not changed. A repeat CT scan (Plate 8(b)) 2 days after the initiation of dexamethasone showed considerable brain oedema around the area of cerebral infarction. Dexamethasone was given for ten days, when a third CT scan (Plate

The first few days

8(c)) showed resolution of the oedema. An angiogram performed on the seventh day after admission demonstrated a complete block of the right internal carotid artery at its origin and narrowing of the left subclavian artery.

The patient made good progress and was independent in the activities of daily living prior to returning home five weeks after admission, at which time little functional recovery had occurred in the hand.

Dexamethasone might also be useful in *preventing* brain oedema but studies designed to answer this question have yet to be done.

Dexamethasone can be given immediately after the acute onset of a 'presumed stroke'. It may be beneficial in cerebral haemorrhage or cerebral tumour but, in any case, can be given before a definitive diagnosis has been made. The general practitioner can give 10 mg of dexamethasone intramuscularly when he first sees the patient. In hospital, a 10-day course can be given, starting at 4 milligrams 6-hourly, reducing the dosage after day 5, or as indicated.

Steroids are not without side-effects. Rarely, gastrointestinal haemorrhage is reported and in some patients abnormalities of glucose metabolism may occur. As with all treatments, the benefits and risks must be weighed before starting, and whilst continuing therapy. A history of peptic ulcer or diabetes mellitus may preclude steroid therapy.

Glycerol

A number of studies in acute cerebral infarction have suggested that glycerol may be useful in reducing brain oedema by exerting an osmotic effect between blood and the brain. Osmosis refers to the passage of a solvent from one fluid compartment to another across a membrane barrier, the direction of flow of the solvent being dependent upon the osmotic gradient between the two fluid compartments. The recommended strength of glycerol is as a 10% solution but the dosage that has been suggested cannot be shown to change the osmotic concentration of the blood to a

The first few days

sufficient degree to cause significant dehydration of brain tissue. Glycerol can also act as an energy source for the brain, although this potential effect has been over-stated in man without objective evidence. The major disadvantage with glycerol is that it has to be given by regular intravenous infusion. The most important side-effect is that it can cause breakdown of red blood cells in the bloodstream (intravascular haemolysis) which might lead to acute shutdown of the kidneys with eventual kidney failure. Glycerol can also exacerbate diabetes mellitus. Further studies are required to assess its role in the treatment of acute cerebral infarction.

Mannitol

Like glycerol, mannitol is also a hyperosmolar agent but, unlike glycerol, it does not readily diffuse into cells. This class of drug suffers from the problem of 'rebound' namely the recurrence of brain oedema when treatment is stopped. The fact that glycerol can diffuse into damaged brain cells means that, potentially, this effect is more likely to occur with glycerol than with mannitol. The action of mannitol lasts less than 4 hours so that repeated infusions are required to prevent 'rebound'. Mannitol, like dexamethasone, is widely used in the treatment of brain oedema following head injury. In our experience, the combined use of mannitol and dexamethasone has been successful in some patients with massive brain oedema (see Case 24). It is also possible that dexamethasone exerts a protective effect on the brain when mannitol therapy is stopped.

Dextran 40

The action of this agent is to reduce the aggregation of platelets and red blood cells in the blood, reducing the viscosity of the blood and preventing platelet emboli. Like glycerol it has to be given by intravenous infusion. In one well-controlled study, dextran 40 was shown to reduce immediate mortality but to have no effect on long-term disability assessed at six months but an earlier study on a small number of patients had shown no clear benefit. A trial combining dexamethasone with dextran 40 in patients with acute cerebral infarction showed no difference between treated and control patient groups.

Vasodilator drugs

A number of agents have been tried to increase blood flow to the damaged area of the brain following cerebral infarction. In 1973 a multi-centre general practitioner trial of isoxsuprine (Duvadilan) was carried out over a 4-month period in the U.K. on 170 patients who had 'mental symptoms suggestive of cerebrovascular insufficiency'. Improvement was recorded in many patients but the trial was uncontrolled, i.e. there were no matched patients who were not given the drug. In an acute stroke trial of patients with presumed cerebral infarction, a different vasodilator, naftidrofuryl (Praxilene), was given up to 6 days after onset and fewer deaths were reported in the naftidrofuryl patients. The diagnosis was not confirmed in these cases and we are repeating this trial in a carefully selected group of patients who have had acute cerebral infarction, confirmed by CT scan.

Vasodilator drugs can be shown to exert an effect on peripheral blood vessels, for example, increasing blood flow to the legs, but it is difficult to show a similar effect on the cerebral arteries since cerebral blood studies are required to prove this contentious point. The most powerful vasodilator of the cerebral arteries is carbon dioxide. The effect of an inhalation mixture of 5% carbon dioxide with 95% oxygen has been used in the treatment of acute cerebral infarction without appreciable effect. Other vasodilator drugs have been tried in the past without clear benefit, although many of the studies have been poorly controlled.

The first few days

Anticoagulants

These drugs have been extensively used for the treatment of acute cerebral infarction and also in the treatment of transient ischaemic attacks. Their use highlights the importance of determining the cause of the stroke as accurately as possible because, if given to a patient with acute cerebral haemorrhage, they will potentiate bleeding with disastrous consequences. Anticoagulants can also cause bleeding into an area of infarction (haemorrhagic infarction).

Anticoagulants can be given to cases of 'stroke-in-evolution' (p. 8) which represent less than 5 per cent of all acute cerebral infarction patients. Typically the patient presents with a minor degree of weakness in the arm or leg which gradually develops into a complete hemiplegia over a period of 24–72 hours. It is not logical to wait until the complete hemiplegia occurs before initiating treatment. The cause of this clinical syndrome is thought to be due to a major thrombus or clot, usually in the internal carotid artery, almost completely occluding the vessel lumen. A thrombosis will form beyond this area of near-occlusion (due to the minimal flow of blood in the distal part of the artery) and extend up the artery towards the brain. This is a very dangerous situation since the thrombus or part of it may occlude major branches of the cerebral hemisphere at any moment.

In order to prevent further thrombosis occurring, anticoagulants are given. They do not dissolve out existing thrombus, but only prevent further extension. Immediate anticoagulant therapy can be achieved using heparin which is given 6-hourly, intravenously, in divided doses. Maintenance therapy is achieved using warfarin, given orally. Special blood tests (clotting time, partial thromboplastin time) are required at regular intervals to monitor the dosage requirements for warfarin (initially daily, gradually every few days, and eventually weekly intervals or longer).

Having started anticoagulants, the problem is to decide how long it is necessary to continue them. This depends

The first few days

upon each individual stroke case and might vary from a 10-day course of heparin only, to 5 years on warfarin following the stroke. Some drugs render warfarin less effective, and others increase the bleeding tendency. Because the list of non-compatible drugs is enormous, great care must be taken about taking other drugs (including alcohol) when the patient is receiving warfarin, and the Haematology Department provides the patient with a list of 'do's and don'ts'. Particular care must be taken with household proprietary medicines containing aspirin. Ideally, no other medication should be prescribed simultaneously.

Anticoagulants may be prescribed in patients with unstable cardiac arrhythmias (e.g. paroxysmal atrial fibrillation) since these patients run a serious risk of an embolus, derived from the heart, lodging in the brain. In patients with proven deep vein thrombosis (p. 88), anticoagulants are started immediately using heparin for 48 hours whilst the oral warfarin begins to take effect. For the *prevention* of deep vein thrombosis, a low dose of heparin injected subcutaneously into the abdominal wall may prove equally effective, as trials of patients undergoing minor (for example hernia operation) or major (for example hip replacement operation) surgery have shown.

Warfarin has been used for nearly 20 years for the treatment of transient ischaemic attacks (TIAs) but a satisfactory answer concerning their precise value in this condition has yet to emerge. Warfarin appears to reduce or even abolish the number of TIAs but it does not seem to protect the patient from a major episode culminating in acute hemiplegia due to cerebral infarction. Antiplatelet agents have received more attention in recent years for the treatment of TIAs.

Drugs affecting platelet behaviour

Exciting advances are currently being made in this new field. Platelets are derived from the bone marrow and are found in the blood in large numbers (200 000–400 000 per

The first few days

cubic millimetre). They are capable of changing shape and aggregating to form a clot. If the stimulus is strong enough, chemicals are released from the platelets which in turn stimulate other platelets to aggregate and release further active substances. This ability to form clots or thrombi is important in preventing bleeding, and patients suffering from diseases where there is a marked reduction in the number of circulating platelets (to less than 10 000 per cubic millimetre) are at serious risk from spontaneous haemorrhage, particularly into the brain. Platelets are to be found in arterial thrombosis but it is not clear whether they have a causative role or whether their presence is secondary,. perhaps as a normal response to vessel wall damage. Clumps of platelets adhering to an area of roughening on an artery wall can fly off into the blood stream (platelet emboli) and are throught to be a major cause of transient ischaemic attacks or 'minor strokes'. Usually the platelet emboli are rapidly dispersed through the arterial tree without permanent consequences, hence the 'transient' nature of the attacks. Increased aggregability of platelets has been demonstrated in a number of conditions, including migraine and stroke. Blood tests designed to show 'platelet hyperaggregability' and, therefore, patients at risk of arterial thrombosis, cardiac or cerebral, have been very difficult to develop because what happens in the test tube (*in vitro*) does not always mimic what happens in the body (*in vivo*).

Three drugs are currently in usage because they modify platelet activities: aspirin (acetylsalicylic acid), 'Persantin' (dipyridamole), and 'Anturan' (sulphinpyrazone). Drugs which can actively boost the body's own mechanisms for inhibiting platelet aggregating are not yet commercially available. Normal vessel walls produce an inhibitory hormone of the prostaglandin type called prostacyclin or PGI_2 which performs this function while platelets produce platelet-aggregating prostaglandins called thromboxanes. Prostacyclin has been synthesized and is currently being tried in different types of arterial thrombosis.

The first few days

Aspirin is widely known for its pain-relieving (analgesic) and anti-inflammatory properties and has been in use for over 50 years. Recently, it has been shown to suppress thromboxane formation and consequently platelet aggregation, but it also inhibits prostacyclin synthesis in the vessel wall, an effect which might actually increase platelet aggregation; this could explain why drug trials with aspirin have been inconclusive.

In 1977, Professor William S. Fields and his colleagues reported the results of a multicentre trial carried out in the United States comparing 650 mg aspirin twice daily with placebo in patients with transient ischaemic attacks, which showed that aspirin could significantly reduce the number of attacks. The Canadian Co-operative study tested aspirin against sulphinpyrazone and placebo and also a combination of the two active agents; after 5½ years of co-operative effort, this was recently completed and reported by Dr. Henry Barnett and his colleagues. It showed no benefit from sulphinpyrazone alone and the combination with aspirin was only slightly better than with aspirin alone. 1 300 mg daily of aspirin, in divided doses, reduced the risk and death rate of further stroke in men but not in women. This sex difference was not shown in the study from the United States but may be important in our understanding of the mechanisms of thrombosis. The dosage of aspirin may also be critical, remembering its different effects on platelet aggregation. Dipyridamole alone has been shown to be ineffective in stroke prevention but combined with aspirin may be of value, and the results of a large-scale multicentre trial will be available in the near future. The role of aspirin immediately following acute cerebral infarction has not been determined although it might conceivably be of benefit.

At present, anti-platelet therapy is reserved for transient ischaemic attacks and prevention of stroke.

Control of hypertension

As we have seen (Chapter 4), hypertension is an important

The first few days

risk factor as a cause of stroke. In many cases, the patient is known to be hypertensive, on or off treatment, prior to the stroke illness, and examination of the retina and heart will provide clues as to its severity and duration. The problem is how best to treat hypertension immediately following an acute stroke, since the blood pressure is very commonly raised *as a result* of this, settling to normal values after 24–48 hours, without treatment. We do not know if this is the reason why a stroke patient with known hypertension and on drug treatment has a raised blood pressure on hospital admission or whether the blood pressure suddenly goes out of control to cause the stroke illness. In the first instance, the same drug treatment might be continued during the acute phase of the stroke illness, but in the second, reduction of the blood pressure might be crucial. How quickly should the blood pressure be reduced and to what levels? A 'normal' blood pressure in a patient with known hypertension would be relatively low and to reduce it to this level could lead to reduced blood supply to cerebral tissue and, if the blood pressure is lowered too quickly, then this could increase an area of cerebral infarction. The danger of too low a blood pressure is illustrated by the following case:

Case 26. A 56-year-old woman was admitted to hospital following a right cerebral hemisphere infarction. She had been treated for hypertension for 8 years and was currently taking treatment to reduce high blood pressure (250 mg methyldopa three times a day and a diuretic tablet). On examination, she had sustained a mild left hemiplegia but the blood pressure was very high, 220/140. Examination of the retina showed haemorrhages and exudates (confirming a severe grade of hypertension) and the heart showed enlargement of the left ventricle (confirming that the hypertension was long-standing). The blood pressure was treated by an intravenous injection of Diazoxide and within 20 minutes the blood pressure was 90/60. One and a half hours later, the patient was drowsy, slightly confused and now had a dense, flaccid left hemiplegia.

Was the presumed extension of this patient's cerebral infarction due to the rapid lowering of the blood pressure

The first few days

to 'below normal' values or was this going to happen despite treatment? A strong argument can be made that it was related to the dramatic fall in the blood pressure. The dose of Diazoxide given to this patient was 100 mg, lower than the normal recommended dose because the patient had been on other blood pressure lowering drugs. Despite this precaution, the blood pressure still fell much too quickly.

The fact remains that adequate drug regimes have not been devised that can lower blood pressure in a controlled and safe way. Intravenous injections carry the risk of being too potent or, if not effective, when repeated, the second dose might potentiate the first. Hydralazine takes longer than diazoxide to work (20 minutes as opposed to 6–8 seconds) and excessive lowering of the blood pressure is much rarer. Probably the best drug available is sodium nitroprusside which can be given by slow intravenous infusion, using a constant infusion pump. It causes a fall in blood pressure within 1–2 minutes, the effect reversing immediately when the infusion is stopped. For this reason, nitroprusside is the safest drug for rapidly lowering blood pressure that is presently available.

There remains the controversy as to what level to reduce the blood pressure. In case 26, we would favour a blood pressure level slightly above normal (e.g. 150/100) rather than 'normal' levels as judged by healthy young adults (120/80 or less).

Supportive therapy

At one end of the spectrum of treatment, this may simply mean trying to cheer the patient up during the traditional, one-hour, hospital visit but at the other end of the spectrum, it means sustaining the life of an unconscious patient, which requires all the skill of the combined medical and nursing team, since the patient is unable to help himself in any way at all. If the patient's swallowing is affected, fluids will have to be given by a feeding (nasogastric) tube or by an intravenous drip. Although a moderate degree of dehydration may

The first few days

actually be beneficial in the first hours after a stroke (to decrease raised intracranial pressure) a decision has to be made concerning the best way to feed the unconscious patient. Fluid requirements and fluid balance are more important than calorific requirements during this stage of the illness. Repeated urinary incontinence may have to be treated by bladder catheterization since 'wet beds' predispose to pressure sores. As a general rule, catheterization should be avoided since prolonged usage inevitably leads to bladder infections; if used, a bladder catheter should be discarded as soon as mobility begins to return. Care of the skin is vitally important and is the reason why nursing staff turn the patient every two hours night and day to prevent pressure sores developing. A sheepskin underblanket for the patient to lie on is particularly useful for preventing bedsores over the sacrum. The comforts and value of sheepskins have been known for centuries and are, in our opinion, more successful than later-day 'ripple mattresses' designed for the same purpose. Positioning of the limbs can be assisted by using extra pillows, for example, to prevent the legs from pressing on each other and to support a paralysed arm.

There is no benefit in prolonged bed rest and it is possible that this can actually predispose to chest infections and thrombosis of the deep veins in the legs, both of which are potentially lethal conditions. The patient can usually sit out of the bed by the third day but this clearly depends upon his clinical condition. Special attention must be taken to position the patient carefully in the chair, supporting the paralysed arm on a pillow and using sheepskin 'booties' for the feet.

Contrary to popular belief, physiotherapy for the paralysed limbs is not required in the acute stage of the illness since the tone in the paralysed muscles is reduced (flaccid) and not yet increased (spastic). Passive movements of the paralysed limbs is best carried out when the patient is being moved and limbs repositioned, but physiotherapy for the chest should be given to prevent the patient becoming 'chesty', a term which does not sufficiently highlight the potential dangers of an acute chest infection.

The first few days

'Supportive therapy' is about what we can do for the patient initially, but it is a tragic mistake to consider him an inanimate subject, which sometimes happens, particularly if speech is affected by the stroke. In the first few days, the patient slowly becomes aware of what has happened to him and realizes that part of his body is not working properly. The medical and nursing team must try to explain to the patient, as simply as possible, that he has had a stroke, why he is in hospital and what the 'team' is trying to do in the way of treatment. It is essential that the patient feels he has a vital part to play in his own recovery, and must be given some idea as to when this will begin. At this stage, it is not necessary to overstress that recovery may be slow and there will be a 'long haul ahead'. Optimism will be created if encouraged by the medical team and if simple goals are achieved. On the other hand there is no place for false optimism or statements such as 'there's nothing much to worry about, you'll be completely better in a week or so' since the contrary will quickly be proved to the patient's consternation and despair. Adapting an old Music Hall adage, 'it's not only what you say but it's the way that you say it'.

10

After the first week

The acute stage of the stroke illness is usually considered to be over after the first week, at the end of which time the stroke patient will often have begun to improve. In particular, he will be aware of his surroundings and consequently a more accurate assessment of speech and intellectual (cognitive) function can be made. Some movement of the leg may be noted at the thigh. Drugs used during the acute phase, e.g. dexamethasone, will have been reduced or stopped altogether, but others will be continued according to the medical indications for each individual patient, e.g. whether hypertension is present; whether anticoagulants have been given in the acute phase and must be continued; changing from an intravenous to an oral regime, as with praxilene. Although the immediate threat to life is over, the patient is still at risk from two important medical conditions, *bronchopneumonia* and *pulmonary embolus* which account for more deaths in stroke patients than actually occur as a result of the acute vascular damage to the brain.

Complications

Bronchopneumonia

In the past, this illness was called the 'old man's friend' because of the rapidity with which it could lead to the patient's demise. Bronchopneumonia is the term used to describe the widespread inflammatory changes that are seen in both lungs as a result of an acute bacterial infection and is the commonest cause of death following a stroke. The diagnosis is based on clinical examination and chest X-ray. The stroke patient's susceptibility is due to the fact that he

After the first week

may be unable to cough properly and that chest expansion is usually poor. A previous history of smoking and chest infections, notably chronic bronchitis, will increase the likelihood that a new chest infection may occur. Preventive treatment consists of physiotherapy to the chest to help expansion and to encourage coughing. Early mobilization and good positioning of the patient in bed or chair also helps. Antibiotic drugs are prescribed according to the type of chest infection. Sputum is sent to the microbiology department for culture of the bacterial organism to determine its sensitivity to different antibiotic drugs. When these results are available, the choice of antibiotics may have to be changed but response to treatment can be dramatic.

In the unconscious patient, there is a very real danger that once a chest infection has occurred it may prove slow to respond. In this situation, the nursing staff carry out frequent suction of the secretions at the back of the throat by means of a narrow bore tube attached to a suction pump, and it may be necessary to administer oxygen. Prophylactic or preventive therapy, administering antibiotics 'in case' the stroke patient might develop a chest infection, is probably unnecessary. Adequate treatment of chest infections occurring in stroke patients would significantly reduce 'stroke mortality'.

Deep vein thrombosis

Any patient, at any age, confined to bed, runs the risk of developing a thrombosis in one or both calf veins. This causes local pain and tenderness in the calf and there may be some swelling of the leg with a slight temperature rise. This complication usually occurs around the tenth day after the stroke but the diagnostic difficulty is that a deep vein thrombosis in the calf can occur silently, that is, without any signs of tenderness or swelling.

The risk of venous thrombosis occurring in a paralysed leg approaches 60 per cent, i.e. it is the rule rather than the exception. Similar high levels of risk occur in patients

After the first week

undergoing orthopaedic operations, for example, fractured femurs. It is common enough after minor surgical procedures involving anaesthetic to be a major concern in post-operative management.

A deep vein thrombosis localized to the calf may, in itself, not be serious but, if the thrombosis spreads up the vein towards the groin, to involve the veins in the pelvis, then there is a very real chance that a major clot may break off into the blood stream. The clot will travel through the right side of the heart and enter the lungs through the pulmonary arteries so that sudden collapse and death will occur if there is a major occlusion of the pulmonary arteries. A number of international multicentre trials have examined the prophylactic use of anticoagulants, usually in small dosage, or of 'dextran 70' in the prevention of pulmonary embolus in hospitalized patients, treatment beginning pre-operatively or immediately after operation. 'Low-dose' heparin, given by subcutaneous injection into the skin over the abdomen, appears particularly useful provided it is started as soon after the stroke as possible. The 'low-dose' regime does not interfere with the clotting mechanism of the body, and so will not cause an increased bleeding tendency or turn an area of brain infarction into an area of haemorrhagic infarction. Heparin will not dissolve out existing clots, but it will prevent a new clot or thrombus from occurring. Simple preventive measurements include elevation of the foot of the bed by 9 inches and the use of elastic stockings.

If a deep vein thrombosis of the leg is suspected, then special investigations are required to confirm the diagnosis, the most accurate of these being phlebography, when an injection of dye is observed in the veins by taking serial X-rays. Treatment consists of anticoagulant therapy, support of the affected leg using a strong 'blue-line' bandage, and bed rest. Once the pain has resolved, the patient should be mobilized as quickly as possible.

After the first week

Pulmonary embolus

A pulmonary embolus is usually a sudden and serious medical emergency, the diagnosis being initially made on the history and examination. It is subsequently confirmed by the electro-cardiogram and also by measurement of enzymes released into the blood because of damage to lung tissue, investigations which will exclude the diagnosis of a possible heart attack as a cause of the patient's sudden collapse. Chest X-ray is not always helpful but an isotope scan of the lung may show areas of lung tissue with a reduced blood supply compatible with the diagnosis. Anticoagulant treatment must be started as soon as the diagnosis is made, in practice, after the electrocardiogram has been taken and before the lung scan is organized and before the results of the blood tests, which will not be available for several days. The problems and risk of anticoagulant treatment have already been discussed more fully (p. 79).

The importance of early mobilization of the patient is that this reduces the likelihood of a deep vein thrombosis in the leg and of a subsequent pulmonary embolus.

Assessments

The need to maintain the patient and institute early medical treatment is the immediate aim of the hospital medical and nursing team. When the drama of the 'life or death' situation has been resolved, then the next aim is accurately to assess the patient and to plan his individual treatment programme. This involves the other members of the hospital team, the clinical psychologist, the speech therapist, physiotherapist, occupational therapist, and social worker. Basically, we want to know 'what the patient can do and can't do'. In order to make our assessments meaningful, a standardized method of examination is required. There is remarkably little agreement as to which of the current methods is best and, in general, it is up to each therapist to choose and sometimes, to devise, her own assessment.

After the first week

Speech assessment

During and immediately after the first week following the stroke, it is important to identify the presence of a speech or communication problem. The speech therapist will provide a diagnosis as to the nature and severity of the speech problem. She should be able to offer practical advice to the medical and nursing staff as how best to communicate, for example: 'use simple, short sentences'; 'the patient understands simple commands'; 'he can understand almost everything, although he cannot formulate words or sentences easily'; 'despite all the words he keeps saying, many of which are nonsensical, he has little insight into his problem and, in fact, understands very little of what you say'. The importance of an accurate speech diagnosis cannot be overstressed.

Case 27. A 67-year-old man was referred to a rehabilitation unit 2 months after his stroke. In the accompanying letter, the referring medical team stated what treatment the patient had received and his subsequent progress. When describing the patient's disability, the letter stated 'dysphasic, unable to say anything or understand; right hemiplegia; emotionally labile and swallowing difficulties'. It was clearly assumed that the site of the stroke was in the left cerebral hemisphere, the probable diagnosis being a left cerebral infarction. The patient had not been previously assessed by a speech therapist.

The patient attended the rehabilitation unit for an initial assessment. He had a residual weakness of the right arm and leg and managed to walk with the aid of one person. The striking feature about him was that he was making guttural sounds but no discernible words. In response to direct commands, he performed the tasks quickly and accurately, even when the tasks became more difficult, e.g. 'show me your left hand; show me your left thumb; put your left thumb on the end of your nose; put your left thumb on your right ear'. This demonstrated that he did not have a receptive problem—he could understand everything! One can imagine the type of indiscretions that might have been voiced by the referring hospital team whilst they were treating the patient, thinking he could not understand anything that was said about him.

This patient was admitted to the rehabilitation unit and this depression (as opposed to 'emotional lability') soon resolved when the rehabilitation team actually talked to him. He was discharged home

After the first week

after a further 6 weeks. At this time, he had persisting dysarthria, or slurring of speech, and was completely independent in all the activities of daily living.

The type of speech problem this patient had is called anarthria (cf. dysarthria) when the patient is unable to produce word sounds because of major articulation problems. The site of the vascular damage is in the brain-stem and not the left cerebral hemisphere and the patient can communicate normally by writing. It is necessary to instruct relatives how to communicate with the patient when there is a speech problem, and this is important for the patient whether in hospital or at home.

Psychological assessment

Psychologists have devised tests to assess brain function, for example memory (long-term, short-term, and immediate recall), left cerebral hemisphere function (verbal tasks), and right cerebral hemisphere function (performance tasks). Clearly, the patient must be able to communicate and concentrate before an accurate assessment can be made. The speech therapist and psychologist are both concerned with communication disorders, although they each assess the problem from different viewpoints. This ability, namely, to be able to look at disability using the expertise of personnel from different medical and allied disciplines, is an important function of the multi-disciplinary rehabilitation team.

The result of the psychological tests are used in the treatment programme, e.g. when planning activities in the physiotherapy or occupational therapy departments. Problems such as visual or sensory inattention may be highlighted, particularly if not detected previously.

Case 28. A 60-year-old woman, who lived alone, was admitted to hospital following the acute onset of weakness involving her left arm and left leg which rapidly improved during the subsequent 10 days. Physiotherapy and occupational therapy was started, initially on the

After the first week

ward, but she was soon going to the therapy departments for treatment. Power in the left arm and leg had virtually returned to normal and physiotherapy treatment was stopped. After 2 months, the patient still was not independent in the activities of daily living. She had problems dressing and performing bimanual tasks, and could not perform adequately or safely in the kitchen. She seemed confused and could not understand why she was doing so badly. She became more and more depressed. The doctors, nurses, and occupational therapist reported to the consultant in charge of her case that 'she was not trying and seemed very depressed'. The consultant suggested a psychiatric opinion. In error, the houseman referred the patient for a psychologist's opinion. The clinical psychologist demonstrated that the patient had a left-sided visual and sensory inattention, so that she tended to ignore the left side of her body and all sensory information that the brain received from this side. The nature of the visual problem was similar to Case 9 (p. 39).

Once this unsuspected disability was identified, the occupation therapy treatment programme was changed and the patient immediately started to improve. The nature of her disability was discussed with the patient and the depression soon resolved with the improvement in accomplishing everyday tasks. Although not completely recovered, the patient was discharged four months after admission, independent in the activities of everyday living. Out-patient hospital follow-up was continued.

This case illustrates the problems that occur if the assessment team (1) does not define the site and nature of the vascular damage from the outset of the stroke illness and (2) does not assess the patient adequately. If it had been realized that this patient had suffered a cerebral hemisphere infarction in the right parietal area, then the medical team might have assessed her differently.

Some patients have a very specific problem in reading and writing which is of particular interest to the psychologist and physician, as in the following case of an architect who could write but not read.

Case 29. A 64-year-old man was treated at home following the sudden onset of weakness in the right arm and leg. After three months, his arm and leg were virtually back to normal. The leg had never been severely affected and his hand was useful after two weeks, although fine movements remained difficult. His family doctor referred him to

After the first week

the outpatient clinic of the neurology department because he was depressed and was not making further progress. The patient could not read and felt totally unable to return to his former employment as an architect.

Examination confirmed the recovery in his right arm and leg. However, when attempting to read, he could see the print on the left half of the page but not on the right side, because of a persisting right homonymous hemianopia (p. 21). A CT scan at this time showed that a left cerebral hemisphere infarction with an area of residual damage in the left occipital area. Although he could see the print on the left half of the page he could not read the symbols. The reason for this was that although the visual information was received in the right occipital area, it could not be relayed to the reading centre in the left cerebral hemisphere because the nerve fibres connecting the two cerebral hemispheres had also been damaged. However, the patient had learnt to overcome this problem partially by tracing the words with the forefinger of his right hand, since the information could then be directly relayed to the left cerebral hemisphere and, in turn, relayed to the reading area. This patient was able to take down a piece of dictation but he was unable to read it back afterwards.

The patient's disability persisted and he was forced to retire. After 18 months, he is beginning to come to terms with his disability, although depression remains a recurring problem.

This syndrome is termed 'dyslexia without agraphia', a technical way of saying that the patient 'cannot read but can write'.

Physical assessment

In order to monitor the patient's progress, he is repeatedly examined on a day-to-day basis by the medical team. More specifically, the doctor must assess (1) the neurological disability resulting from the stroke and (2) the presence of any disability present before the stroke that might hinder rehabilitation and recovery.

1. *Motor system.* The amount of weakness in the arm and leg is ascertained according to the major muscle groups. The arm and leg may be equally affected or the arm may be more affected than the leg; rarely, the leg is more affected than the

After the first week

arm. Although gross movements at the shoulder may be possible, hand movements are usually severely affected at the outset. In the leg, movement may be seen in relation to the thigh, but marked weakness at the ankle causes the affected foot to drop which can cause walking difficulties. Weakness can be measured on a 0–5 scale: 0 = no movement and 5 = normal power.

It is also important to assess the tone of the limb, i.e. whether the arm is more floppy than usual (flaccid) or stiffer than usual (spastic). Some degree of spasticity is usually required in the leg before the patient can begin to walk.

Incoordination without much weakness may be seen after a stroke affecting the brain-stem. Following a cerebral hemisphere stroke, slight weakness of the limbs can mimic incoordination because the patient cannot perform precise or fine movements. Only after a long period of disuse does the patient lose muscle bulk. The reflexes are tested by the doctor using a 'patella hammer'. When the tone in the limb is reduced (flaccid) the reflexes are reduced or absent, but are very brisk when spasticity is present. The physiotherapist is very experienced in assessing motor function in terms of power and tone, and combined assessments with the doctor are of considerable value when monitoring physical recovery.

2. *Sensory system.* The sensations of pain and temperature are conveyed by the same pathway of sensory nerve fibres to the brain. Light touch and joint sense are conveyed by a different route. Both pathways converge in the brain-stem before being relayed to the parietal area in the cerebral hemisphere. As a result, a stroke in the cerebral hemisphere involving the parietal area may result in loss of sensation of the opposite side of the body.

Sensory inattention has already been discussed (p. 37). This is much more common when the right parietal area is affected than the left, and the consequences of missing this diagnosis have been illustrated (Case 28, p. 92).

After the first week

3. *Visual system.* The occipital cortex receives visual information from the opposite visual field. Strokes affecting the back of the brain may cause visual field disturbances only. Visual inattention is also an important clinical sign, occasionally missed (Case 9, p. 39). When a visual field defect is present this should be properly chartered using simple techniques (perimetry) available in the eye department.

4. *Other disabilities.* After taking into account the previous medical history, full examination of the patient will reveal other causes of disability not related to the acute stroke illness, and apart from residual disability due to a previous stroke. Major causes include: poor visual acuity due to cataracts; diabetic disease affecting the eyes; reduced mobility because of arthritis or orthopaedic problems; other neurological diseases, such as Parkinson's disease or dementia; vascular disease, such as ischaemic heart disease and peripheral vascular disease affecting the legs, both of which can seriously limit exercise tolerance. The presence of co-existing causes of disability will not only affect the patient's rate of progress during rehabilitation but will delay eventual hospital discharge. The following case illustrates the dangers of bed rest.

Case 30. An 84-year-old woman had been severely affected by arthritis for 12 years, her knee joints being grossly enlarged, painful, and swollen. Despite this, she managed to get around her small flat although she could not venture out and relied on daily visits from her daughter who had a family and was employed full-time. Although she had a home-help and meals-on-wheels, 5 days a week, she fended for herself at weekends. After the sudden onset of a right hemiplegia one evening, she was admitted to the local geriatric unit. Three weeks later, the patient appeared to have made little progress and a neurologist's opinion was sought.

On examination he found that her vision and speech had never been affected and there was no residual neurological deficit attributable to the stroke but, following three weeks in bed, she was immobile due to the arthritis in her knees and hips. Despite all her courage (she was a charming and intelligent old lady), her legs did not improve and after 8 months in hospital, arrangements were made for her to be transferred to a long-term geriatric ward.

After the first week

Social assessment

It is necessary to obtain as much information as possible concerning the social circumstances of the stroke patient, by talking to him, the immediate family, relatives and neighbours visiting the patient in hospital. An elderly person, living alone at the time of the stroke, with only a neighbour cited as next of kin in the nurses' daily report on the ward, is likely to need far more practical help prior to, and following, hospital discharge. If the 'social network' for each patient is defined as soon as possible after admission, then arrangements can be made to help the patient well in advance of the proposed discharge date. In 1976, we devised a simple method for portraying the social network in the form of a diagram (Fig. 8, p. 111).

Case 31. A nonagenarian's social support. A 91-year-old woman was admitted to hospital following a right cerebral infarction causing a left hemiplegia. She had been known to be diabetic for six years, receiving insulin injections given daily by the district nurse. Three weeks after admission she was transferred to a rehabilitation unit and was discharged home 3 weeks later. At this time, she was able to walk using a frame but had no functional recovery in her left arm. At home, she continued to progress despite arthritis of the left hip, and one year later she could wash and dress herself, and could walk slowly without an aid. Discharge home would not have been possible without the close support of all her family. She lived alone on the ground floor of a two-storey terraced house owned by her son-in-law. The patient's daughter and the daughter's family lived upstairs. Although out during the day, they were able to provide support during the evenings, nights, and at weekends. During the day, another daughter stayed with her. The district nurse called in daily.

This case illustrates a particularly close family and social network which made early discharge from hospital possible for this elderly patient. The social network diagram shows the lines of communication that exist and this can be used as a guide when considering discharge home. Solid lines mean daily visits. New lines of communication can be added, for example, visits from a district nurse or home-help. A daily visit by a member of the family or community health team is essential when the patient lives alone.

11

After the first month

At this stage some patients have recovered sufficiently either to return home or to be transferred to a rehabilitation unit. However, the rate of recovery is variable, as discussed in Chapter 15. The stroke patient who stays in hospital may be moved from the acute end of the 'Florence Nightingale' type of ward, and receive less medical attention. He still requires a great deal of nursing care, physiotherapy, and occupational therapy but the total amount of care will vary according to the number of staff available. In a busy district hospital, with few physiotherapy staff, it is possible that each stroke patient on the ward will receive less than half an hour of individual treatment during each week day, and less time or none at all during the weekends.

Depression accompanies stroke in at least 70 per cent of cases. This can act as a real barrier to progress since it will dampen the patient's attitude towards therapy which in turn will discourage the nurses and therapists from trying too hard. Depression can be mingled with frustration, demonstrated by throwing food on the floor, not talking to visitors and so on. In a few instances, specific drug treatment in the form of antidepressants may be helpful, but the patient really needs time and a lot of attention from caring staff and relatives. The importance of family involvement, if at all possible, has already been stressed.

The aim, by the end of the first month, is for the patient to walk, with aids if necessary. Following an acute stroke, walking is the single most important activity and arm recovery should be considered a bonus at this stage. In an analysis of stroke patients, we found that 9 out of 10 patients were able to walk before discharge. Put another way, if the patient cannot walk by himself, then the chances are remote that he

After the first month

will achieve a sufficient level of independence to enable him to return home. Minor speech problems (usually dysarthria) and visual difficulties may have spontaneously resolved. If the rate of recovery up to this point has been rapid, then sometimes this may seem to slow down or even 'plateau' after the first month before further change occurs.

Motivation is the most difficult factor to assess and to alter and can affect predictions of outcome, favourably or adversely. The personality of the patient prior to the stroke is clearly important. Previous life-style, social circumstances, and the way that the patient has coped with previous crises may provide important clues to the patient's attitude towards medical nursing and paramedical staff. When the patient appears not to be motivated, it is essential for his rehabilitation that at least one member of the hospital team should try to gain his confidence and talk to him, discuss his fears for the future, and reassure him.

Rehabilitation

A stroke patient is often in hospital for a long period of time, more than 3 months in some instances. The duration of his stay in hospital depends upon the degree of physical disability caused by the stroke illness, the presence of other non-related disabilities, and the patient's social network, i.e. who is at home, the home conditions, and the availability of family and friends for support. Many doctors believe that rehabilitation begins as soon as the patient is admitted to hospital and, in a sense, this is true, but in practice nurses are too busy and short staffed, and physiotherapy sessions on the ward often last minutes rather than hours.

The hospital ward is hardly the ideal environment for the stroke patient, and the longer the patient stays on the ward, the less 'acute' he becomes in medical terminology, and consequently the less interest may be shown by the medical team in his progress. Medical wards are considered to be acute admitting wards and, in fact, the majority of patients

After the first month

stay in hospital for less than 3 weeks. There is no adequate provision for patients who stay in hospital for 1-3 months, what has been termed 'intermediate stay'. At this stage, patients require a high degree of nursing and paramedical care and not much in the way of an acute medical service orientated towards urgent investigations and acute medical treatment. A 50-year-old patient who has just had a heart attack is given a long time to recuperate before he resumes his normal activities, so that it is optimistic to expect a 70-year-old stroke patient to recover completely from his illness and return home if nothing had happened within one month.

The rehabilitation unit may appear to be the answer but in this country, there are very few units and recently, with financial cutbacks, several well-established units have been closed or threatened with closure. In addition, rehabilitation medicine is not yet a recognized speciality in the United Kingdom whereas in the United States, for example, large hospitals specializing in rehabilitation have been created. Patients attending rehabilitation units tend to be one of two types: those recovering from an orthopaedic operation (e.g. hip replacement) or those recovering from a neurological illness causing disability, such as stroke or multiple sclerosis. Although the turnover of orthopaedic patients is usually quicker, more neurological patients occupy the beds in the rehabilitation unit at any one time. Of these, stroke patients are by far the largest group. There are very few specialist rehabilitation units of the type found at Stoke Mandeville (paraplegics) and the Wolfson Rehabilitation Centre at the Atkinson Morley's Hospital, Wimbledon.

The criteria for entry to a rehabilitation unit also vary and the first consideration is often whether there is a rehabilitation unit within the region. It is unlikely that the stroke patient will be considered for admission to a rehabilitation unit by the medical team at the district general hospital when the rehabilitation unit is 30 miles away, and the family may turn down the opportunity because of the difficulties of visiting. Clearly, both these factors, although important, are

After the first month

not at all related to the individual needs of any particular stroke patient.

Although referral to a rehabilitation unit is made by the consultant in charge of the patient, the rehabilitation unit team like to assess the patient before considering admission. If a patient is turned down because his disability is still too severe, he returns to hospital until further spontaneous recovery takes place when re-referral might be tried. The psychological effect on the stroke patient after he has gone through this performance can be imagined. Some referring doctors have even suggested that many patients have to be of 'Olympic standard' before they are admitted to a rehabilitation unit. Alternatively, a place may be offered in 1, 2 or 3 months' time rather than straightaway.

There is an important psychological time in the recovery pattern of any stroke patient when a different environment, a different medical team, and a different rehabilitation approach can be very stimulating.

The rehabilitation unit

The size of different units varies enormously, from a capacity of just over 20 patients to more than 60 patients. There are usually gymnasium facilities for physiotherapy, 'light' and 'heavy' workshops run by the occupational therapists, and rooms for the speech therapist. Some units employ a remedial gymnast in addition to physiotherapists.

The fabric of the unit may be less important than the concept of the team approach to rehabilitation. The weekly round-table conference is the most important meeting of the team, at which all members are present, medical, nursing, and paramedical. The 'chair' is taken by the consultant in charge of the unit and each of the patients is discussed in turn. Each member of the team involved in the patient's treatment programme provides a progress report, and the following week's treatment is planned. These weekly assessments are included in the medical case notes, providing

After the first month

a valuable record of the patient's progress. A card showing the treatment timetable can be given to the patient each week and this is a useful way to show the patient what is planned for him, reinforcing what he has already been told. The card can also be shown to family and relatives when they visit. A daily treatment programme consists of not less than 3 hours and usually 4 hours or more, divided into morning and afternoon sessions. Activities, i.e. physiotherapy, occupational therapy, and speech therapy, are varied according to the patient's needs.

The high level of activity expected in the rehabilitation unit means that many stroke patients are never admitted, because they are not capable of sustaining the necessary intensity of work. Hence there is a need for an intermediate type of unit from which patients could graduate to longer periods of activity, if necessary.

The duration in the unit is very variable but most patients stay 4–6 weeks. If progress is not made, the patient is referred back to the district general hospital.

Every rehabilitation unit is different, differentiated partly by the personnel involved in patient care and partly by the medical interests of the consultant-in-charge. The patient and his family will get to know each and every member of the rehabilitation team.

12

Surgical treatment of stroke

Surgery has an important part to play in the prevention of stroke but, until recently, it was thought to have no place in the treatment of acute stroke or during the months immediately after a stroke. Surgical attempts to remove atheromatous debris and clot from a freshly blocked internal carotid artery had proved to be disastrous, often making the stroke patient far worse, and such attempts were abandoned for more conservative medical treatment. Recent advances in surgical techniques have opened up a new era in vascular surgery of the brain.

Stroke prevention

Patients with transient ischaemic attacks (Chapter 2) require urgent investigations to discover the cause of the attacks, the most important investigation being angiography (Chapter 6). If extensive atheroma is found in one or other of the common or internal carotid arteries, a direct surgical approach to remove the atheromatous material can be considered. If the atheroma is more widespread, then surgery is likely to be contra-indicated in favour of medical treatment, mainly anti-platelet drugs (p. 80). When atheroma is confirmed to the neck arteries on the side referrable to the patient's symptoms and signs, then surgery is probably the treatment of choice. To remove an area of atheroma, the surgeon has to strip the inner lining of the artery, a procedure called carotid endarterectomy which can be done with little risk.

If a neck artery has been blocked for some time, however endarterectomy may actually be dangerous since, during the procedure, emboli may fly off into the brain causing an acute

Surgical treatment of stroke

cerebral infarction. The blocked artery can sometimes be bypassed by means of a shunt.

Surgery following stroke

The introduction of the operating microscope has meant that vascular surgeons and neurosurgeons can perform very precise, delicate surgery which had not been previously possible. The skilled surgeon can now join together (anastomose) arteries which have a diameter of between 0.75 mm and 1.5 mm. Extracranial to intracranial arterial anastomoses were introduced because the results of surgery on completely blocked internal carotid arteries were so poor. The aim of operation is to bypass the block in the internal carotid artery to provide more blood to the cerebral hemisphere. Since the external carotid artery, the other branch of the common carotid artery, is rarely blocked by atheroma, one of its branches on the temple (the superficial temporal artery) is used. A bone disc is raised on the temple (about 8 cm in diameter) and branches of the middle cerebral artery are inspected on the surface of the brain. The largest vessel seen is chosen for the other end of the anastomosis and the superficial temporal artery is sewn to the side of this vessel. The bone disc is replaced and patency of the anastomosis can be demonstrated by pulsation over the superficial artery, by ultrasound techniques (directional doppler), or angiography. The operation is technically straightforward in skilled hands, and is performed either by a neurosurgeon or a vascular surgeon working with a neurosurgeon.

Two questions need to be answered about the place of this operation in the management of stroke patients: (i) Why should this operation benefit stroke patients at all? and (ii) Which patients should be offered this operation? The answer to the first question has not been fully explained. Traditional teaching has been that following acute cerebral infarction, no recovery of nerve cells or fibres can occur amongst the infarcted (dead) tissue. The area immediately surrounding

Surgical treatment of stroke

the infarcted tissue is initially affected by the acute vascular disturbance and physical recovery is thought to be related to recovery in this region. Why then, if the blood supply to this region is increased following this operation, should some patients improve when their acute stroke illness had occurred several months before and their degree of disability had remained static? The suggestion has been made that even several months after the stroke illness, some cells around the region of the infarcted nerve tissue remain viable, but not properly functioning because of inadequate blood supply. Increasing the blood supply to these nerve cells could therefore improve their functioning and this is the rationale behind the operation.

It is much more difficult to decide which patients should have the operation. Patients with cerebral haemorrhage are clearly excluded. The operation has been advocated in patients with transient ischaemic attacks, reversible ischaemic neurological deficit, or cerebral infarction with moderate disability. A large international multicentre trial based in Canada has been set up to compare patients who have had the operation and those who have not. Except in carefully selected patients, it is important to await the results of this study before the operation becomes too widely used in an unselected fashion; otherwise popular demand for the operation might seriously affect a proper evaluation.

At present the operation of extracranial to intracranial arterial anastomosis should be considered in any patient who fulfils the following criteria: (1) he has had an acute cerebral infarction (demonstrated by CT scan); (2) the angiogram demonstrates a complete occlusion of the neck artery (common or internal carotid) on the same side; (3) there is a moderate degree of disability (memory slightly affected, slight dysphasia, weakness primarily of arm); (4) the acute stroke illness occurred 6 weeks previously; (5) there is no evidence of heart disease. These recommendations are probably stricter than some experts might quote and no doubt they will require revision in the years to come,

Surgical treatment of stroke

particularly when the results of the Canadian study are available.

Whilst there is still little place for surgery in the acute management of stroke, cerebral, and particularly cerebellar, haematomas can sometimes be removed during the first few days.

13

The patient at home

The exact number of patients staying at home after a stroke-like illness is not known for certain, but is in the region of 40–50 per cent. The decision to send a patient to hospital does not necessarily relate to the severity of the stroke (see Chapter 8), and it cannot be assumed that 'mild cases' stay at home and 'severe cases' go to hospital. Although social and environmental factors are important, the ultimate decision depends upon the general practitioner and what he *believes* to be best for the patient. This book has so far presented 'the facts' as seen from the viewpoint of the hospitalized patient. The dangers inherent in making generalizations about 'hospital care', as though there was a set package of treatment for stroke, are obvious enough. When discussing 'home care', it is even more difficult to generalize. From visiting stroke patients at home, our experience has ranged from those happy meetings when an interested and committed family look after 'mother or grandmother' who has had a stroke to tragic cases at the opposite end of the scale.

Case 35. A 67-year-old man had developed a mild right hemiparesis following a 'presumed stroke'. He had persisting difficulty with his right hand and dragged his right foot and saw his general practitioner 3 days later. No home visit was made nor any attempt at formal assessment, physical or social.

Three weeks later, neighbours contacted the general practitioner in a panic because somebody had gone into the patient's home for the first time, possibly in 20 years. The general practitioner's findings were horrific. The patient lived in terrible conditions—a small terraced house, only one room occupied by the patient, the rest being filled with rubbish. Rain regularly came through the roof, and the patient's lavatory had not worked for months. The passage way leading to his room was filled with foul, stale water. His bed linen and mattress were saturated

The patient at home

with urine. He had not eaten properly for over a week; before then a neighbour had given him food over the fence, but had since gone on holiday. The patient could not go out shopping, prepare food for himself, or cope in any way, and so was admitted to hospital. During the last 6 months of his hospital stay, he had made a complete recovery, but since a temporary placement was unacceptable to the patient, he could only be discharged after new housing was found, and this took 9 months.

At the initial interview with this patient, his general practitioner had asked 'do you think you can manage?', and the answer 'yes, I think so', really meant 'I'm really not sure'. A home visit in these circumstances determines the true situation. The patient was reluctant to leave his home in spite of the appalling conditions. The local social services department sent their 'heavy brigade', a team of cleaners who have the job of sorting out homes in this plight, but on this particular day, two or three members of the team were on holiday and the remaining member of the 'heavy brigade' refused to do the job. The director of housing could not believe her report, and sent another assessor, whose report was even worse. An outside contractor was eventually commissioned to do the work. All this took 9 months to declare that the property was unsuitable for human habitation, which was really known from the time of the first visit! In the meantime, the patient had to wait in hospital, 'blocking a bed', at a cost to the taxpayer of over £500 a week. Early discharge would have been possible if the patient had accepted alternative accommodation sooner.

Patient care at home can be considered under the following headings: (1) medical problems; (2) nursing problems, and (3) support services; (4) the patient who has been discharged from hospital.

Medical problems

As we have seen (Chapter 5), it is not possible to determine the type of stroke without special investigations (Chapter 6). The best diagnosis that can be made under these circumstances is 'presumed stroke'. In addition disturbances of cardiac rhythm or blood pressure cannot be monitored regularly during the acute stage of the illness outside hospital.

The patient at home

Nursing problems

The spouse or family at home are unlikely to have relevant experience. After a severe stroke, it may be very difficult to give the patient a drink or deal with incontinence problems. Keeping the patient in bed, with all its attendant problems, is often easier than trying to mobilize the patient early by encouraging him to sit out in a chair. The problems associated with transferring the patient onto a commode, to a lavatory in the house, or an outside lavatory, may reach insuperable proportions and this alone could necessitate hospital admission. In less serious cases, the spouse may cope admirably but will still need guidance on how best to mobilize the patient. Help from visiting therapists or attendance at the nearby hospital for outpatient rehabilitation may be required.

Counselling can be a major problem, as illustrated in this next case.

Case 36. Another 60-year-old taxi driver had been visited by his general practitioner after sustaining a mild left hemiparesis following a presumed right cerebral hemisphere infarction. Initially, he made good progress but at 6 weeks found that he still had some weakness of the left hand. He became very depressed because he thought he would never be able to drive again. As a result he decided to sell his cab. After 3 months he was seen by a specialist who predicted a favourable outcome with further recovery of his hand function. In the meantime, he had used the proceeds of the sale of his cab for housekeeping and had not approached the social services department for financial help. At 9 months, he had fully recovered from the effects of the stroke and was back in full-time employment as a taxi driver. The use of his financial capital to replace his loss of earnings meant that he was unable to buy a new cab, so he had to hire one.

Nursing problems at home should never be underestimated when the question of hospital or home care is discussed with the family. The resilience of many spouses and families can defy description and their self-sacrifice and uncomplaining devotion often go unrecognized, especially when the main concern may be with failing district social services and

The patient at home

patients alone at home. The social revolution which followed the end of the Second World War has had important implications for the care of the elderly at home. Communities have been irrevocably broken up in many instances with all the attendant problems imposed because families and neighbours are no longer in a position to help each other.

Support services

Very few people know the name and address of their local Social Services Office. In many instances, they are unaware of the facilities that can be provided and to which they are legally entitled. The general practitioner can ask his health visitor and district nurse to see the patient regularly. The district nurse can help with bathing the patient, transferring the patient from bed to commode in the morning, and, if necessary, putting the patient back to bed later in the day. Aids, such as a commode, can be ordered through the Social Services Department. Re-housing can be carried out only rarely but the area occupational therapist can visit the patient at home, assess his needs, and advise the Social Services Department about adaptations to the home which may be necessary; these have to be implemented by law. Since this work can take time to carry out, an early visit by the occupational therapist is recommended.

Until recently, no district or area physiotherapists or speech therapists were available to make house visits, but approaches should be made to see if this is possible in the patient's home area. The social work department in the hospital and in the community work closely together. A visit from a social worker can be of considerable help since she can inform the patient and his spouse about what is available in the way of help and what they are entitled to in the way of benefits. They can also develop a 'social network' for the patient living alone at home with a 'hot-line' if the situation breaks down.

The patient at home

Case 37. A 76-year-old woman had suffered three episodes of speech loss due to a left cerebral hemisphere infarction. After each of the two previous episodes her speech had recovered completely within 6 weeks but, following the third episode, her speech was severely affected. She could use the telephone to state her name but was able to say a few words only. She managed at home and was independent in the activities of daily living but was too frightened to go out.

The social network diagram (Fig. 8) shows the patient's lines of support and communication. Daily visits were made by the caretaker and a friendly shopkeeper would send a delivery boy with the shopping. The social worker helped the patient to arrange her financial affairs.

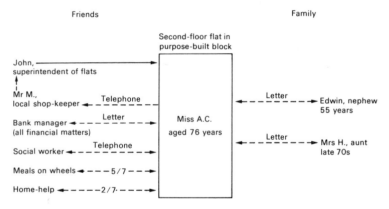

Fig. 8. An example of a social network diagram. The solid line means daily visits; the dotted line less frequent communication and visits; $x/7$ = the number of days per week these take place.

In our opinion, home assessment by these different members of the community health team are essential. Referral to the hospital outpatient department should be considered as soon as possible after the patient is mobile and, despite the problems of ambulance transport, the patient can be assessed and treated in the therapy departments on an out-patient basis. Treatment should be planned to take place at the peak of the recovery phase during the first 3 months to provide

The patient at home

'bursts' of daily intensive therapy rather than two or three half-hour visits a week, which is possibly of little value (Case 32, p. 121). The spouse should be encouraged to enjoy the break that these hospital visits afford but his or her attendance at some of the therapy sessions is important in order to show precisely what the patient can and cannot do.

Consideration should also be given to the possibility of the patient attending a Day Centre, where he can be given physiotherapy. These centres provide an important community function since they focus health care in a special centre. The medical co-ordinator for the patient's care at home is his general practitioner, who can make a joint assessment of the patient at home with any of the district or area therapists, so that short term and long term planning can be a 'team' decision.

Home care following hospital discharge

Much of the benefit that might have accrued from hospital in-patient treatment can be spoilt by early or precipitous discharge from hospital. This happens particularly when there has been inadequate communication with the spouse and family, so that little or even no preparation is made at home prior to discharge.

Case 38. 'The first thing I heard was when my 11-year-old daughter went after school to visit my husband in hospital and was told by the ward sister to tell me he was to be discharged. She came straight home and, 2 hours later my husband arrived by ambulance. I had to bring the bed down to the front room because he couldn't climb stairs, and it has stayed downstairs ever since. He had to use a plastic bucket as a toilet for the first few nights until a commode came'.

A home visit should be arranged by the hospital team well in advance of discharge. Ideally the social worker and occupational therapist carry out a joint visit so that they can see the home and assess the situation, bearing in mind the patient's disabilities. Sometimes it is possible to take the patient on the visit for a few hours to see how he will actually

The patient at home

cope and to see how the spouse or family will manage. There is a lot of difference in being able to manage the lavatory and bath in the occupational therapy department of the hospital and carrying out the same activities in the home, where facilities for disabled people are usually less than ideal. The assessors must be experienced.

Case 39. A 64-year-old woman was married to an 81-year-old World War I veteran. She had suffered a dense left hemiplegia but had recovered sufficiently to walk with an aid and her husband's help. Their housing, although sub-standard and due for demolition, was considered adequate prior to hospital discharge as it was a ground floor flat. At home, the patient was very depressed and lost all interest in her walking. A wheelchair was ordered by the hospital team so that her husband could take her out locally. At the entrance of the flats, there was a 12-inch step, an obstacle which was impossible for the husband to negotiate with the patient in the wheelchair. This point was missed by the assessment team at the time of their domiciliary visit, and this patient had been housebound for a year.

Reports to the general practitioner should be sent before hospital discharge and arrangements should be made for a doctor from the hospital team to see the patient at least once in the hospital outpatient department. If possible, this doctor should also talk to the general practitioner on the telephone before the patient is discharged to provide details of the patient's disabilities, support at home, and domiciliary services that have been ordered. For follow-up purposes, it is important to define medical responsibility clearly since, all too often, the general practitioner still considers the patient to be the responsibility of the hospital when the next outpatient appointment is in fact in 6 months' time or the appointments have been stopped, whilst the hospital doctor considers the patient to be entirely under the care of the general practitioner.

Long-term outlook

The future can be full of uncertainties for both the patient

The patient at home

who has never been admitted to hospital and for the patient discharged home after hospital treatment.

It has been estimated that 50 per cent of all stroke patients may not survive the first year after their stroke illness. This percentage is predominantly made up of patients dying within the first 3 months from secondary complications, bronchopneumonia or pulmonary embolus, the major causes of death following stroke, or from the acute effects of the stroke itself.

The pattern of recovery is as follows: (1) the patient first learns to walk again, (2) speech recovers to about 60 per cent of the patient's former ability in most instances, and finally, (3) the arm recovers, but this can be very variable. Recovery occurs in step-wise fashion with periods of 'plateauing' during which frustration and depression are common. Much debate has been given to whether left cerebral hemisphere strokes have a better or worse prognosis than right cerebral hemisphere strokes. Speech disabilities can be very severe (left hemisphere function) and spatial problems, notably neglect of the left side of the body (p. 37) can be severely incapacitating (right hemisphere function). The site, size, and nature of the stroke are the final determining factors relating to long-term outlook, as discussed in the next chapter. Recovery certainly occurs for up to 2 years and, in our experience, further recovery can occur even *after* this length of time.

It is important for the patient to make a psychological recovery even in the presence of persisting disability. Depression occurs because of a number of inter-related factors. These are:

(1) *restricted social activity* as a result of the stroke (this also includes patients who do not yet feel 'back to normal' because of memory or minor speech problems, although otherwise fully independent);

(2) *loss of work*, particularly if the stroke causes premature retirement;

(3) *loss of independence* – patients may not be able to do even the simplest of tasks;

The patient at home

(4) *feeling less 'well-off'* due to loss of earnings;
(5) *an awareness of being 'at risk'*, particularly when the stroke patient is living alone;
(6) *dissatisfaction with housing*;
(7) *changed family role* – the patient may be unable to adjust to or accept 'role-reversal' with the spouse;
(8) *problems in the marital relationship.*

In a few patients, the stroke may exacerbate unresolved feelings in relation to retirement or recent bereavement. Many of these patients have been depressed before the stroke or had a previous psychiatric history.

Whenever possible, domiciliary follow-up visits should be made by the appropriate members of either the community or hospital team, especially to patients who are severely disabled and likely to be, or known to be, at risk. Follow-up by the primary care team and the social services is necessary since additional problems at home may arise because of the development of other illnesses or increasing age of the patient and his family. Regular follow-up should help to prevent acute crises. From the medical point of view, prevention of a further stroke and heart disease (Chapter 4) and the proper treatment of hypertension and diabetes are the goals to aim for.

Consideration should also be given to further physical therapy and even admission to a Rehabilitation Unit, if possible, as indicated by the stroke patient's recovery. Holidays for the patient and his wife or the patient alone can be arranged by the Social Services Department.

In our opinion, a better understanding between hospital and community-based teams can be achieved; this would markedly improve the quality of life for the stroke patient and his family.

14

Assessment of the treatment of stroke

In view of the large number of new stroke cases that occur each year, one might expect that medical and surgical treatment regimes have been evaluated without too much difficulty, but this has not yet been done. During the last 15 years, a number of drug trials have been carried out in different parts of the world to assess the effect of new drugs in the acute treatment of stroke, but nearly all of these trials suffer from serious faults in methodology. It is virtually impossible to make meaningful comparisons between stroke patients receiving different treatments when the following factors remain uncontrolled: selection of patients for referral to hospital, timing of treatment, diagnostic accuracy, site and extent of brain damage, disability, ratings, previous medical history, and social history. We shall now look at each of these factors in turn, remembering that for a valid comparison of any two methods of treatment, all other factors should remain constant.

There is no consistency in the pattern of stroke patient referral to hospital (Chapter 7), partly because hospital-based trials have not shown clear benefit to stroke patients as a whole or to any subgroup in particular.

In drug trials, it is important to start treatment at the same stage in the evolution of the illness. Many stroke trials have compared patients who have received the active drug 4 hours after the onset of the stroke with those receiving it after 4 days. It is clearly too much to expect that any drug, given in the same dosage, will be equally efficacious at any point in an illness.

As we have seen in Chapter 5, there are many different types of stroke. It is quite wrong to lump them together as if they were a homogeneous group since each type of stroke

Assessment of the treatment of stroke

has a different natural history. It would be valid to study cases of cerebral infarction given different treatments, but investigations should be required to define the patient groups as accurately as possible.

Stroke affecting different areas of the brain causes different patterns of disability, and it is logical to compare groups of patients who have all suffered, for example, an infarction in the left cerebral hemisphere, and have similar levels of disability. (The size of the brain damage may be seen on the CT scan although we can partly assess this indirectly from the amount of disability present.) Most stroke studies have not considered this degree of matching, and have not compared patients with similar degrees of disability from the outset and have not followed the patient's progress for a long enough period.

In order to measure recovery, all stroke studies use some type of disability rating score. It is often difficult to know exactly how different hospital teams have carried out their assessments (medical, speech, physiotherapy, etc.). As a result, a prerequisite to any new stroke study is for the hospital team to develop their own rating system. This means that, in most instances, no two studies can ever be strictly compared because of the differences in methodology and, in particular, assessments.

Outlook or prognosis following a stroke is obviously affected by the patient's previous medical history, for example, if he has had a stroke before, or is known to be hypertensive or have heart disease. The presence of these and other illnesses will affect rehabilitation and recovery.

Even when all other factors are equal, the date of hospital discharge will vary according to home circumstances, for example who is available to look after the patient and the type of housing (for example, bungalow or thirteenth-floor flat with poor lift services), as much as to the efficacy of treatment.

Assessment of the treatment of stroke

Stroke trials

We shall now look at how the trials were designed and what conclusions can be drawn from them.

Virtually all have been hospital based. Before entry to any trial the patient or a relative has to provide 'written, informed consent', in other words, that they understand the purposes of the trial; that they are willing to participate; that they understand the patient may be in the 'placebo' (no treatment) group, and that the study has been passed by an independent hospital ethical committee.

The trials have either compared an active agent with a placebo or two different active agents. Some studies have taken a consecutive group of stroke patients admitted to hospital without attempting to make a specific diagnosis and randomly allocating patients to 'treatment' and 'placebo' groups.

Most studies have tried to adopt a 'double-blind' procedure, that is, neither the patient nor the treating doctor know whether the patient is receiving 'active' or 'placebo' treatment, thereby reducing the degree of observer bias. Studies have been carried out in patients thought to have cerebral infarction only or cerebral haemorrhage, mainly based on lumbar puncture findings (p. 50) so that there has been great variation between the studies. As a result, some studies have ensured that treatment must be started within 24 hours of the stroke, whilst in others, up to 5 days or even longer has been allowed. The number of patients involved in each study has varied enormously, as has the period of observation, 2 weeks to one year. Each centre has used a different neurological or disability rating scale. Most studies have matched the patients well for age, sex, and 'neurological score', but there have been notable exceptions.

The conclusions drawn from these studies are contradictory, for example, dexamethasone, a steroid, is said to be 'beneficial in stroke' by one report, to have 'no effect in cerebral infarction and haemorrhage' in another; a third study found 'no indication for its use in stroke'. In view of

Assessment of the treatment of stroke

the major differences in design between these studies, these conflicting results are hardly surprising.

The generally accepted conclusion amongst physicians is that no single drug has been shown to be of benefit for the treatment of stroke, but our view is that a carefully designed drug trial with particular attention to the problems of matching patients has yet to be carried out. Until these problems in design are overcome with treatments using a single drug, it will be even more difficult to evaluate multiple drug therapy for acute stroke which is advocated, although 'not proven', by medical colleagues in some countries.

Remedial therapy

The number of therapists attached to any hospital is decided by the availability of funds, and most departments seldom achieve their full complement of staff. In district general hospitals, there is a chronic shortage of therapists, coping with inadequate facilities and equipment. The best departments are to be found where a school is attached for one of the therapy specialities, e.g. physiotherapy, because early in their training, students are involved in clinical work, and can help the qualified staff. The larger departments of physiotherapy can be superbly equipped with, for example a large gym, workshops for occupational therapy, and a hydrotherapy pool. In general, most hospitals have some physiotherapy service but very limited occupational and speech therapy services. In recent years, each of these specialities has been concerned with upgrading their training course, even to degree level. At a time when departments of physical medicine began to decline, they have also sought autonomy. The source of patients for treatment depends upon an adequate referral service from either the hospital doctor for in-patient work or the general practitioner by out-patient referrals. Treatment may also be continued on patients discharged from hospital, and some centres provide an out-patient rehabilitation service.

Assessment of the treatment of stroke

Considering the vital role of therapists in the rehabilitation team, it may seem unthinkable to question the value of their work. Doctors in charge of such teams or those working closely with therapists inherently believe that 'therapy works'. Other doctors, less interested in rehabilitation and the paramedical problems, may not refer a patient for therapy ('What's the point? They can't help this patient').

Does therapy work? The majority of therapists are young women, enthusiastic about their work and eager to help patients. Because of their vocational interest the level of motivation and co-operation of the stroke patient is increased. Is the patient's improvement a direct result of this interest or has the therapist's treatment actually worked? It is very difficult to test 'therapy treatment'. New drug treatments, for example, can be tested by controlled clinical trials—by comparing active treatment against placebo, then assessing the efficiency of the new drug against the 'standard' or 'accepted' drugs. To do a controlled trial of a particular therapy then the equivalent of a 'placebo' group would have to be a 'no-treatment' group and many therapists shrink from this idea. In a situation where therapists are paid only by their patients, as in North America, setting up such a trial would be impossible.

The fact is that (with one exception) there have been no adequately controlled trials of remedial therapy—physiotherapy, occupational therapy, or speech therapy. As a result there is no proven 'standard' or 'accepted' treatment in any of these professions. Therapists, although realizing this basic truth, have been slow to accept this challenge—a challenge which would help define useful treatments as well as those patient groups likely to benefit from treatment. A therapist often 'instinctively knows' which patients to treat and which patients may not be helped by treatment. Sometimes they have difficulty in quantifying progress—the patient may say he is better and the therapist may think he is better, but the improvement is qualitative, and therefore difficult to measure.

Assessment of the treatment of stroke

In practice, the duration of therapy treatment in hours bears little relationship to the type of stroke and speed of recovery. Discharging a patient because of a lack in improvement may be a difficult moral decision; the therapist knows that no other physical treatment will be offered as an alternative. Reduced treatment regimes, e.g. twice-weekly out-patient occupational therapy, may be impractical and of no benefit to the patient, as illustrated by the following case history.

Case 32. A 61-year-old man, who lived alone, was attending out-patient occupational therapy twice a week following a right cerebral hemisphere infarction causing predominantly weakness of the left arm. Eleven out-patient visits were booked. Ambulance transport failed to come on three occasions. The total treatment time for the eight visits was 3½ hours. The round trip, home to hospital out-patients to home, was 8 hours.

When planning a trial, the first major problem is to design accurate, simple, but useful assessment techniques which can be quantified and tested, comparing the results obtained by different assessors (inter-observer variation) and repeated assessments on a static subject at different times (intra-observer variation). A two-point scale of the type: (1) disability absent and (2) disability present, has a high inter-observer correlation approaching 100 per cent, but lacks sensitivity. Introducing a three-point scale (e.g. (1) disability absent, (2) disability slight, (3) disability present) increases sensitivity slightly at the expense of an increased inter-observer variation. When a four- or five-point scale is used (for example left arm (1) not affected, (2) mildly affected, (3) moderately affected, (4) severely affected) inter-observer agreement can be as low as 75 per cent.

For this reason we have looked for objective methods of measurement rather than relying only on clinical rating scales. For stroke patients, we have used gait charts to examine the walking problems of different subjects using bioengineering equipment (polarized light goniometer). Interestingly, the

Assessment of the treatment of stroke

charts obtained demonstrate that even the 'normal' leg behaves abnormally after a stroke.

The single most important trial of out-patient rehabilitation following stroke has been carried out at Northwick Park Hospital. It took over 6 years to recruit just more than 100 patients. Of all the patients referred to the trial, just over 10 per cent were able to satisfy the stringent criteria for inclusion. As a result, most of the patients recruited to the trial were in the 'mild-to-moderate severity' category. Patients were randomized into three treatment groups: (1) those receiving intensive therapy (4 full days a week), (2) those receiving 'standard' therapy (3 half-days a week), and (3) those receiving no therapy (controls). All the stroke patients were carefully assessed at regular intervals and patients in the 'no-therapy' group were visited regularly at home. Therapy varied in intensity (4 days, 3 half-days, none) rather than in the type of treatment. Patients receiving treatment had physiotherapy and occupational and speech therapy according to their needs. The period of follow-up was for 1 year.

The results of the trial showed that intensive therapy was significantly better than no treatment at all but no statistical difference could be shown between intensive and 'standard' therapy and 'standard' therapy and no therapy. An oversimplified recommendation based on these results would be 'intensive therapy or none at all'.

Similar studies employing different trial designs are urgently required. Certainly our experience is that therapists can help stroke patients during periods of spontaneous recovery. They have little or no success when plateaux (i.e. no change) occur along the recovery graph. At this point, frustrated patients want *more* treatment, not less. Until trials of therapy are carried out, the treatment given will continue to be more of an art than a science.

In the last few years, a new scheme has been tested which involves an interested but untrained body of helpers, namely volunteers. Miss Valerie Eaton Griffith organized a project for stroke patients with severe speech disorders (dysphasia) in

Assessment of the treatment of stroke

her community. Seven lay members were recruited for every single dysphasic stroke. All were women with no previous experience in this work, and none of the women had been acquainted with the patient before his illness. They each took turns on a daily basis to visit the patient. At first, the helper would do most of the talking but, little by little, the patient would join in more, and from there progressed to playing word games (Scrabble) or cards. Outings were planned and even simple outdoor activities that the patients had lacked confidence in doing (such as shopping). Gradually the patients were encouraged to meet each other and join clubs.

This project was so successful that it has been extended to different areas each with a co-ordinator linked to the patient's general practitioner and the area community physician. The project was sponsored by the Chest, Heart and Stroke Association, a registered charity, and helpers received no expenses for their efforts. Of considerable importance was that this treatment was free—to the patient and to the National Health Service, relying on the goodwill of the helpers and the interest of the patient and his family. The result of the project demonstrates at least one important point—the proper evaluation of therapy, whether it be physiotherapy, occupational therapy, or speech therapy, should begin without further delay.

In spite of the problems that have been encountered in research into the treatment of stroke in the past, there is every indication that the carefully controlled trials now being set up will yield results that will provide a national basis for treatment in the future.

15

Outcome after stroke

Predicting the outcome, or prognosis, following a stroke is only possible if the course (natural history) of the illness is known. Statistics based on an analysis of 'stroke patients' with no attempt to differentiate the type of stroke illness have little meaning for the individual patient, because each type has a different natural history and prognosis. The most important factors affecting outcome are (1) the *site* of damage in the brain, (2) the *size* or extent of the stroke, (3) the *nature* of the stroke (i.e. precise diagnosis). Other factors include (4) the rate of onset of the stroke, (5) intellectual ability and personality before the stroke, (6) age, (7) sex, (8) general health, and (9) social circumstances.

Site of the stroke

Strokes affect the cerebral hemispheres more commonly than the brain-stem, and hemisphere strokes are possibly more common on the left than on the right. As a rule, patients who have had brain-stem strokes and who survive the initial episode have a lower mortality. The mortality from brain-stem strokes depends more upon the nature of the stroke (i.e. haemorrhage or infarction) than in the case of cerebral hemisphere strokes; eventual recovery is also better in patients who have had brain-stem strokes.

Extent of brain damage

A large area of brain damage will be associated with a greater degree of neurological disability and is more likely to be life-threatening. The extent of the brain damage can be ascertained initially from the clinical signs present and subsequently

Outcome after stroke

confirmed by a CT scan. The presence of hemiplegia, hemianaesthesia, and homonymous hemianopia means that there is a large area of damage in the part of the brain supplied by the middle cerebral artery. If damage on the left side is responsible, speech may also be affected. Hemiplegia or hemianaesthesia alone are more likely to be due to damage to a small branch vessel, or a vascular lesion in the internal capsule (p. 21) where the nerve fibres from the cerebral hemispheres descend in a narrow band. Homonymous hemianopia occurs after occlusion of one of the posterior or middle cerebral arteries.

Nature of the stroke

In the cerebral hemispheres, *infarction* accounts for over 85 per cent of cases, the remaining being due to *haemorrhage*. The mortality within the first month of infarction lies between 25 and 40 per cent. In nearly half the cases, cerebral infarction has occurred secondarily to an embolus, either from an atheromatous plaque in the common or internal carotid artery in the neck (75 per cent of cases) or from the heart (25 per cent of cases). In addition to the extent of the brain damage produced by the embolus, the prognosis will also depend on the extent of atheroma elsewhere in the body and if heart disease is responsible for the embolus. The remaining cases of cerebral infarction are secondary to thrombosis or occlusion of a major blood vessel in the neck (carotid or vertebral). As a rule, cerebral infarction due to an embolus has a better prognosis than when due to thrombosis, since emboli tend to 'break up' within the cerebral arterial tree and disperse. As a result, some patients may make a dramatic recovery a few days after sustaining a dense hemiplegia.

Internal carotid artery occlusion can occur without causing any neurological deficit, especially if it occurs gradually. This is because an adequate collateral blood supply may be developed through the circle of Willis (p. 11). Sudden occlusion of the internal carotid artery produces a large area of infarction in the cerebral hemisphere. A similar picture is

Outcome after stroke

seen when the first part of the middle cerebral artery, which is the direct continuation of the internal artery, is occluded. Occlusion of the middle cerebral artery is nearly always caused by an embolism. An occlusion further along the middle cerebral artery causes a smaller area of infarction, mainly affecting the face and arm.

Infarction in the brain stem has a lower mortality than infarction in the cerebral hemisphere, being approximately 25 per cent. Survivors also do much better, with most patients making a complete recovery. If the area of infarction occurs in the upper part of the brain stem (the pons) then the prognosis is very much worse; many of these patients die within a few days, because vital centres controlling the heart, circulation, and respiration are situated in this region.

Haemorrhage into the cerebral hemispheres has a greater mortality than infarction. The haemorrhage may be localized within the substance of the brain and can resolve spontaneously with little residual neurological deficit. Haemorrhage rupturing on to the surface of the hemispheres or into the ventricles of the brain (where the cerebrospinal fluid is formed) is associated with rapid loss of consciousness and a very high mortality, over 80 per cent. Paradoxically, survivors may have a better degree of recovery after cerebral haemorrhage than after infarction, since blood may track between nerve fibres without destroying them. Haemorrhage into the brain stem carries a very serious prognosis and if the bleed is into the upper brain stem (pons) it is almost invariably fatal. Haemorrhage into the cerebellum has a variable prognosis but is amenable to surgical drainage.

Rate of onset

Sudden onset of stroke is most commonly seen following a cerebral embolus. If the embolus quickly breaks up, then very little permanent neurological deficit results. If a thrombus or clot develops behind the embolus, occluding a cerebral artery, then the resulting brain infarction is less likely to be

Outcome after stroke

reversible. If the embolus then suddenly moves on, the infarction may also become haemorrhagic. Haemorrhage into the brain can produce signs swiftly but usually the effect is maximal after 3 hours. For the purposes of diagnosis, however, the rate of onset of the stroke has little value. When the onset is associated with rapid loss of consciousness then the prognosis for immediate survival is very poor.

Intellectual ability

Recovery of speech and language function is certainly related to the patient's intellectual abilities before the stroke.

Case 33. A 72-year-old man, a distinguished physicist, suddenly lost his speech and became confused and disorientated. On examination, he was found to have dysphasia, predominantly expressive or non-fluent in type. Investigations confirmed a small localized left cerebral hemisphere infarction.

His native language was Hungarian but he had lived in Great Britain for more than half his life. Prior to his stroke he was able to speak five languages fluently, his other languages being German, Spanish and Italian. The first language to recover was English. Only after 3 months did his Hungarian return to conversational standard and he did not regain his other languages. Total recovery of language function was only 60 per cent compared with his previous abilities, and this prevented him from going on a planned lecture tour.

Recovery of language in a polyglot is of considerable interest, since the first language returning gives us some idea as to what the patient naturally feels is or has become his native language. Clearly, English and not his native Hungarian had become this professor's first language.

Case 34. A 64-year-old Polish man, a head waiter, had been in Great Britain for over 30 years. He suffered a left cerebral hemisphere infarction which caused a non-fluent dysphasia, weakness of the right side of the face, and of the right arm. He had always lived alone, did not live in a Polish community and, at work, spoke English most of the time. The patient's speech began to return after 10 days; after a further two weeks, his Polish had improved although it was not back to normal, but his English never returned to his former level of performance and remained very poor.

Outcome after stroke

Recovery of reading is probably related, in part, to previous ability. Memory and concentration do not always improve and may remain a persistent problem, even after full recovery of any limb weakness.

Age

Mortality from stroke increases with age, but this may be because, first, other serious illnesses are present which delay recovery and secondly, older patients are more susceptible to complications such as bronchopneumonia and pulmonary embolism. Many older patients over 70 years of age can make good recoveries so that the patient's age should not militate against him when he is being considered for referral for investigation and treatment. Strokes occurring in children or young adults have a better prognosis, often with good recovery, even after severe neurological disability at the onset.

Sex

Women are more likely to survive to old age than men and consequently are more likely to suffer a stroke. There is no evidence to suggest that there is any difference in the rate of recovery following stroke between men and women.

General health

The presence of high blood pressure and heart disease will affect the long-term prognosis of stroke patients. After a stroke illness, patients making a full recovery run a greater risk of having a heart attack than a further stroke. Poorly controlled blood pressure, uncontrolled diabetes mellitus, abnormal concentration of fats in the blood stream, and also excessive viscosity of the blood predispose to either heart disease or stroke. A past history of chest disease, such as chronic bronchitis and emphysema, may make the patient

Outcome after stroke

more likely to develop a chest infection after the stroke. Arthritis and orthopaedic problems may delay rehabilitation (p. 96).

Prediction of immediate outcome

It is not possible to diagnose accurately the nature of the stroke from the history and clinical examination alone, particularly as massive cerebral infarction and cerebral haemorrhage cause similar clinical pictures. Several studies, including our own, have shown that the immediate outcome following stroke can be predicted according to the level of consciousness after the stroke; this is the single most important prognostic sign.

In patients who remain fully conscious and alert over the initial 24–48 hours, the average mortality is far less than the 25 per cent which is the average mortality for all infarctions. If the level of consciousness is lower, and the patient is sleepy or drowsy most of the time, then the mortality rate more than doubles to 50 per cent. In those patients who are more deeply unconscious, although they may react purposively to external stimuli, the rate increases to 75 per cent. A patient who remains deeply unconscious for much longer than 36 hours has little chance of improving and virtually all patients in this group die.

Two other signs present in the first 24 hours correlate with a poor subsequent outcome. If the patient's arm and leg have total, flaccid paralysis, then there is little likelihood of a full physical recovery. In addition, the patient may not spontaneously move his eyes towards his paralysed limbs, because the centre for conjugate eye movements has been damaged; if this sign persists after 4 days, there is likely to be a poor physical recovery.

Prognostic accuracy will be increased if the nature of the stroke has been ascertained. A deeply unconscious patient who has had a cerebral hemisphere haemorrhage, as demonstrated by CT scanning, or a pontine haemorrhage, has a poor

Outcome after stroke

prognosis and is unlikely to survive 7 days. A patient who has sustained a small cerebral hemisphere infarction, as shown on CT scan, who has never had any alteration of consciousness, and has only a mild hemiparesis, is likely to have a good prognosis for full physical recovery providing he does not develop pneumonia or a pulmonary embolus.

Predicting long-term recovery

The over-all mortality for 'stroke' 1 week after the acute onset of the illness is 30 per cent. Of the survivors, 20 per cent will make a complete recovery and a further 10 per cent will have 'minor' disability; disability in the remainder varies between moderate to severe.

The long-term recovery also depends upon the site, size, and nature of the stroke. If the affected limbs remain flaccid, the outlook is worse, but increase of limb stiffness (spasticity) appearing in the arm and leg within the first week is a good prognostic sign. If spasticity has not become evident in the leg at the end of 1 month, then the chances for the patient being able to walk again are poor. Frequently the arm remains flaccid although the leg becomes spastic; these patients learn to walk but their arm remains functionally useless. Recovery in both the arm and leg begins with muscles nearer the trunk, that is to say, at the shoulder and in the thigh. In the leg, lifting of the foot at the ankle (dorsiflexion) is an important movement for recovery because it means that the patient will be able to raise his foot without stubbing his toes whilst relearning to walk. Many patients begin to manage this movement while the other muscle groups in the thigh and leg are beginning to get stronger. Arm function takes very much longer to return and recovery is often incomplete. Fine movements of the thumb and fingers are the last activities to return, and may take months or even up to 2 years. It is important for the shoulder to be kept mobile, if necessary, by passive exercises, the patient using his strong arm to exercise his weak arm. If a 'frozen' shoulder develops it may

Outcome after stroke

seriously delay recovery in the affected arm.

Surprisingly, homonymous hemianopia does not present a great handicap to rehabilitation since most patients quickly learn to overcome this visual field defect.

The presence or absence of these neurological signs depend upon the site and size of the stroke and the rate of recovery depends both upon these two factors and the nature of the stroke illness itself.

16

The future

Prevention of a stroke

The statistics show a decline in the incidence of stroke due to haemorrhage and infarction, and this engenders a good deal of hope for the future. If better control of high blood pressure has produced this result, then more effort must be put into the early detection and treatment of hypertensive subjects.

The importance of transient ischaemic attacks as a warning sign of an impending stroke is well known to specialists in this field, but more attention to these sort of attacks must come from other doctors. The diagnosis in these cases, unlike that of stroke, is based on the history and clinical findings (recovery within 24 hours) but all these patients should be referred to the hospital specialist for further assessment. It is hoped that the development of 'non-invasive' techniques will improve and reduce the need for performing angiography when assessing arterial disease in the neck arteries, but it will be still required for the investigation of damage to arteries inside the brain.

Despite the lack of evidence to implicate conclusively diabetes, lipids, and smoking as definite risk factors, patients should be aware of the advantages of good control of diabetes, dietary restriction for raised blood fats and should seriously consider giving up smoking since ischaemic heart disease is a definite risk factor (Chapter 4).

Medical and surgical treatment of transient ischaemic attacks has undergone a minor revolution which shows no sign of abating. Medical treatment consists of antiplatelet therapy, although the best drug is yet to be determined. Surgical treatment was originally confined to carotid endar-

The future

terectomy (i.e. the clot and narrowing within the artery was removed in those patients with unilateral carotid artery disease producing symptoms), but extracranial-intracranial arterial anastomosis is now being done in a few centres for this type of vascular disease (Chapter 12). Research is currently being carried out in many centres throughout the world looking at each of the aspects discussed above.

Acute treatment of stroke

There is, as yet, no adequately designed trial of stroke treatment in the medical literature but new possibilities for drug therapy exist, for example prostacyclin (p. 81). Whilst it is possible that a single drug regime may be unsuccessful in the treatment of the acute stroke illness, the stroke patient should, if possible, be given the opportunity of hospital treatment during the acute phase of the illness. In Europe and United States, up to 80 per cent of patients are admitted within 24 hours, whilst the equivalent figure in the United Kingdom is 40 per cent. Until this situation improves, acute therapy will not be available to the majority of patients.

Research is being directed towards finding a drug that will increase blood supply to the brain and be given to the patient immediately on admission to hospital. If possible, the general practitioner should have the same drug available to him, so that once the diagnosis of 'presumed stroke' has been made, the drug can be given, intramuscularly or intravenously, if necessary.

There is a need for new well-designed multi-centre stroke trials, using the CT scanner. Controlled trials of therapy are urgently required, and particular attention should be paid to the number of patients in the trial and the trial design.

A number of stroke units should be set up in the U.K. to reduce secondary complications and lengthy hospital stays. The work and the results of such units should be compared with similar patients admitted to non-specialist hospital wards.

The future

Consideration should be given to the provision of more day centres rather than more rehabilitation units, but each region in the country should have at least one rehabilitation unit as well as a unit for disabled patients who are not of geriatric age (*young chronic sick* unit).

Surgery of stroke

We have entered a very exciting new phase of cerebrovascular surgery, but it is imperative that the new technique of extravascular to intravascular arterial anastomosis (see Chapter 12) should be assessed adequately before it is widely performed.

Glossary

acute: of sudden onset, reaching a peak within 24 hours.
aneurysm: an abnormal sac-like swelling arising from the wall of an artery, often where it branches.
aorta: the main artery leading from the heart.
aphasia: literally 'without speech'. The loss or impairment of the capacity to use words or symbols for ideas. Aphasia is caused by damage to the dominant hemisphere, the left hemisphere in right-handed people. *Expressive* aphasia (dysphasia): inability to speak; *receptive* aphasia (dysphasia): inability to understand the spoken word. Note: *aphasia* and *dysphasia* are frequently used synonymously.
apoplexy: an anachronistic term for stroke.
arteriosclerosis: literally 'hardening of the arteries': the process by which fats and lipids, often with calcium, are deposited in the arteries, causing thickening of the walls and loss of elasticity. In severe cases, arteries became narrowed and may become blocked.
atheroma: the name given to the 'porridge'-like material which is deposited in the arteries in arteriosclerosis.
basilar artery: supplies blood to the brain-stem and the back of the cerebral hemispheres, especially the visual area. The artery is formed from the joining together of the two vertebral arteries.
brain scan: a synthetic radio-isotope is injected into a vein and the appearance of the isotope in the brain is measured by a gamma camera. This procedure gives considerably less information than the CT scan, which has superseded it.
brain-stem: the name given to the part of the brain that connects the cerebral hemispheres and the cerebellum to the spinal cord. (In the brain-stem, the upper motor neurones cross over, so that the motor area of the left cerebral hemisphere controls the right arm and leg, whilst the right cerebral hemisphere controls the left side of the body.)
cervical vertebrae: The bones of the neck. The vertebral arteries, right and left, travel upwards within bony canals in the cervical vertebrae before joining to form the basilar artery. Bony degeneration of the cervical vertebrae may cause impairment of blood flowing in these arteries (vertebrobasilar ischaemia).
cerebellum: the part of the brain that lies below the cerebral hemispheres but above the brain-stem and is primarily concerned with coordination of movement.

Glossary

cerebral hemispheres: the largest portion of the brain, occupying the whole of the upper part of the skull and consisting of right and left hemispheres, each of which have specialized functions (see Chapter 3).

carotid artery: one of the major arteries arising from the aorta. The right and left carotid arteries divide in the neck into the external carotid artery, which supplies the face and scalp, and the internal carotid artery, which supplies the cerebral hemisphere. Both carotid arteries and the single basilar artery are united at the base of the brain (to form the circle of Willis).

craniotomy: the operation of opening the skull to expose the brain.

cerebral embolus: (*see also* **embolus**): this blocks an artery. Cerebral emboli may come from atheromatous plaques in the internal carotid artery or from the heart, causing transient ischaemic attacks or cerebral infarction.

cerebral oedema: swelling of the brain due to damage of brain cells following brain infarction. The swelling is due to accumulation of fluid (oedema) within and around the cells.

cerebrovascular accident: this term is often used synonymously with stroke but is not a diagnosis.

cerebrovascular disease: the term used to cover all types of vascular disease of the brain, including stroke.

cerebral haemorrhage: bleeding into the substance of the brain tissue. Poorly controlled high blood pressure is the commonest cause. This type of stroke is seen in 10–15 per cent of cases.

cerebral infarction: death of brain tissue. It is the commonest type of stroke, about 85 per cent of all cases. Approximately half of the cases are due to emboli, the other half to thrombosis.

cerebral thrombosis: the occlusion of a blood vessel by clot (*see also* **thrombus**).

circle of Willis: named after Thomas Willis (1621–75), who described the circle of arteries which communicate at the base of the brain, an arrangement that allows for cross-flow to an area of brain when its own blood flow is reduced.

completed stroke: a stroke that has reached its maximum level of disability.

computerized axial tomography (CAT or CT scan): an X-ray procedure requiring a computerized scanner which gives cross-sectional views through the brain at different levels. This technique will differentiate strokes caused by haemorrhage from those due to infarction. The procedure carries no risk to the patient, but the equipment is expensive and confined to a few centres.

dysarthria: impairment of articulation of speech.

electroencephalography (EEG): using special equipment with electrodes and amplifiers, the electrical activity of the brain can be monitored and the print-out or permanent record interpreted.

Glossary

embolus: a piece of matter in the bloodstream, usually a clot, which is made up of blood constituents. Other types of emboli occur, e.g. fat, air.

hemianaesthesia: loss of sensation affecting the arm and leg on the same side of the body.

hemiparesis: weakness of the arm and leg on the same side of the body, e.g. right hemiparesis means weakness of the right arm and right leg.

hemiplegia: paralysis or complete weakness of the arm and leg on the same side of the body; more severe than hemiparesis.

homonymous hemianopia: the loss of vision in half the visual field (hemianopia) on the same side of each eye (homonymous). A right homonymous hemianopia is loss of vision in the right (outer) half of the right eye and in the right (inner) half of the left eye.

hypertension: raised blood pressure; above normal values for age. Patients with poorly controlled high blood pressure are at risk from heart disease and stroke.

ischaemia: reduced blood flow.

lumbar puncture: in this procedure a needle is inserted into the back to obtain cerebrospinal fluid for examination of cells, protein, etc.

prognosis: the predicted or expected outcome.

reversible ischaemic neurological deficit (RIND): an acute disturbance of brain function of vascular origin causing disability lasting more than 24 hours but resolving, often within one week, always within one month.

stroke: an acute disturbance of brain function of vascular origin causing disability lasting more than, or death within, 24 hours.

stroke-in-evolution (evolving stroke): the term used to describe a stroke when the maximum level of disability has not yet occurred.

transient ischaemic attack (TIA): an acute disturbance of brain function of vascular origin causing disability which lasts *less* than 24 hours. Most TIAs last less than 6 hours, and many are fleeting.

thrombus: a clot of blood formed in a blood vessel which may block an artery supplying blood to the brain.

vertebrobasilar ischaemia: reduced blood flow in the vertebral and basilar arteries, often associated with bony degeneration in the cervical vertebrae.

Index

acalculia 19
acetylsalicylic acid 80, 81, 82
acute, definition 5
acute stages of stroke 87
    management 73-86, 108, 133
age
    incidence, related to 25, 27
    prognosis, related to 128
agraphia 16, 93-4
aids 63, 98, 110
alcohol 30
anarthria 92
aneurysm 7, 44-5, 54
angiography 12, 57-8, 103, 132
    non-invasive 58
antibiotics 88
anticoagulants 79-80, 87, 89, 90
antidepressants 98
Anturan 81
aorta 23, 37
aphasia 18, 65, 66
apoplexy 10
arterial anastomoses 10, 104, 134
arteries of brain 10-12, 21 (*see also* *specific arteries*)
arteriosclerosis 7, 22-4, 25, 54
arteritis 32
aspirin 80, 81, 82
assessment
    patient 90-7
    treatment 116-23
atheroma 22-4, 29, 37, 103, 125
atherosclerosis 7, 22-4, 25, 54
axons 21

bed rest, dangers 85, 88, 90, 96
bed sores 61, 85
behaviour, antisocial 61, 98
blood cells, red 77
blood clot 7, 45, 79, 81, 89, 126
blood flow, cerebral 10, 11, 15, 20, 40
    vasodilators 78
blood pressure 6, 7, 23, 24, 83-4, 128
blood supply 5, 6, 7, 10-12, 105, 125
    to brain stem 8
blood tests 53, 72, 79, 81, 90
brain
    atlas 13, 14
    damage 6, 12, 16-22, 32, 35, 124-5
    functions, maps of 12-22
    left side, stroke affecting 16-22, 34-6, 72, 93, 125
    lobes of 17-22
    right side, stroke affecting 16-22, 37-9, 124
    scan 55-6 (*see also* computerized tomography)
    'silent' areas 14
    swelling, *see* cerebral oedema
brain-stem 8, 11, 21-2, 42, 92, 124
    haemorrhage 43-4, 126
    incoordination 95
    infarction 41, 126
Broca's speech area 16, 18
bronchopneumonia 87-8

calculation 17, 19
carbon dioxide 78
cardiac arrhythmias 80
carotid arteries 10, 11, 12, 23, 32, 103, 125
    bruit 40, 42
    calcifications 54
    stenosis 37, 56
    surgery 104, 105
    thrombosis 79
catheterization 85
CAT scanning 17, 50, 51, 56-7, 72
causative factors, of stroke 22-4 (*see also* risk factors)

Index

cerebellum 11, 22, 126
cerebral arteries 11, 12, 21, 78, 126
cerebral embolus 7 (*see also* embolus)
cerebral function, disturbance of 5
cerebral haemorrhage 5, 7, 24, 25, 33
   brain swelling 74
   differential diagnosis 45, 50-2, 55, 57
   drug treatment 76
   prognosis 25, 33, 125-6, 129
cerebral hemispheres 6, 8, 11, 13, 34-9, 72, 124
   dominant, left 14, 16
   interrelationship 15, 20
cerebral infarction 5, 7, 12, 23, 34, 40, 125
   brain swelling 74
   differential diagnosis 50-2, 55, 57
   drug treatment 74, 77, 78, 79, 82
   outcome 125
   surgical treatment 104-6
cerebral oedema 35, 55, 57, 73, 74
   drug treatment 75-7
cerebral thrombosis 7, 46
cerebrospinal fluid (CSF), 43, 44, 50-1
cerebrovascular accident (CVA), 5-6
cerebrovascular disease 5, 6
cervical vertebra 11, 42-3
chest infection 62, 85, 87-8
children, strokes in 32
cholesterol 23, 29
cigarette smoking 30
circle of Willis 10, 11, 125
clots 7, 45, 79, 81, 89, 126
communication, *see* speech
'completed stroke' 8
complications 87-90
computerized tomography 17, 50, 51, 56-7, 72
consciousness 33, 35, 41, 73, 74, 127, 129
constructional ability 19, 39
contraceptive pill 27, 46
coronary arteries 23, 29, 30, 36-7
coronary care unit 59
cortex 15, 21
counselling 109
craniotomy 13

creatine phosphokinase (CPK) 53
CT scan 17, 50, 51, 56-7, 72
CVA 5-6

day centre 112, 134
deep vein thrombosis 80, 85, 88-9, 90
depression 98, 114-15
dexamethasone 75-6, 87, 118
dextran-40 77-8
dextran-70 89
diabetes mellitus 29-30, 76, 77, 115, 128
diagnosis of stroke 16, 50, 72 (*see also* investigations)
Diazoxide 84
dipyridamole 81, 82
disabilities 2, 3, 5, 16, 26, 92, 94-6, 117 (*see also specific disabilities*)
   prediction 12
'disturbance of cerebral function' 5
doctors
   role of 60-1
   strokes in 2
Doppler effect 58
double-blind trial procedure 118
dressing 19, 39
drop attacks 8
drug treatment 74-82, 87, 132, 133 (*see also specific drugs*)
   trials 116-19
Duvadilan 78
dysarthria 65, 66
dyslexia 19
   without agraphia 94
dysphasia 18, 71, 123

echo-encephalography (EchoEG), 51-2, 72
electrocardiogram (ECG) 30, 37, 53, 90
electroencephalography (EEG) 45, 54, 72
embolus 7, 23, 36, 77, 80, 103
   platelet 81
   prognosis 125, 126
   pulmonary 87, 89, 90
emergencies, medical 69-70
emotion, changes in 16, 61, 98

Index

endarterectomy 23, 57, 103
enzymes 53, 90
'evolving stroke' 8
exercise 31
expressive dysphasia 18
extracranial to intracranial arterial anastomoses 104-6, 134
eye movement 16, 17, 35, 129

facial muscles 22
family, care by 3, 61-2, 66, 97, 98
see also home, care at
fear 2, 3
feeding difficulties 61, 64, 85
first-aid 71
fissure of Rolando 18, 19
fluid requirements 85
focal neurological deficit 71
frontal lobe 14, 16-17

glycerol 76-7
grey matter 21, 57
growths, *see* tumours

haematoma, subdural 45-6, 55
haemorrhage
    brain-stem 43-4
    cerebral, *see* cerebral haemorrhage
    subarachnoid 32, 44-5, 51
haemorrhagic infarction 51, 79, 89, 127
handedness 16
heart attack 26, 36-7, 53, 59
heart disease 25, 27, 29, 30, 32, 115, 128
hemi-anaesthesia 125, 130
hemianopia 20, 21
hemiplegia 5, 6, 9, 35, 71, 125, 130
    in children 32
heparin 79, 80
    'low-dose' 89
heredity 28
history, medical 70, 71
home, care at 3, 68, 97, 107-15
    after hospital discharge 112-13
    longterm outlook 113-15
    medical problems 108
    nursing problems 109
    support systems 110-12

homonymous hemianopia 20, 21, 35, 125, 130, 131
hospital 107
    admission to 26, 68-72
    discharge from 112-13
    intermediate stay in 100
    monitoring progress 73-4
    rehabilitation 98, 99-100
    specialist personnel in 59-67, 86, 90, 98, 119
hydralazine 84
hyperlipidaemia 29
hypertension 6, 7, 23, 24, 28, 29
    control of 82-4, 115
    stroke prevention 31

incidence of stroke 25-6, 32
incontinence 85, 109
incoordination 22, 41, 95
infarction
    cerebral, *see* cerebral infarction
    haemorrhagic 51, 79, 89, 127
    myocardial 26, 36, 39
intellectual ability 127-8
internal capsule 21, 22
investigations 50-8, 72
ischaemia 5, 6, 8, 42-3
ischaemic heart disease 25, 27
isotopes 55-6
isoxsuprine 78

language 127 (*see also* speech)
lesion, site of 13, 16-22, 71-2, 93, 124
leukaemia 47
limb, tone of 85, 95, 129, 130
lipids 23, 29, 128
lumbar puncture 50-1, 72
mannitol 77
memory 17, 19, 92, 128
meningism 45
meningitis 32, 45
migraine, hemiplegic 48
mobilization, early 61, 62, 85, 88, 90, 109
monitoring, patient progress 73-4
mortality rates 26, 27, 33, 114, 124, 125-31
motivation 99, 120

Index

motor area 15, 17, 18
motor neurones 20-2
motor system, assessment 94-5
movements
    control 5-6, 18, 19, 20, 22, 130
    physiotherapy and 62, 65
muscles 22, 65, 85, 95, 129, 130
myocardial infarction 26, 36, 39

naftidrofuryl 78, 87
nursing care
    at home 109-10
    in hospital 84-6, 98-9
nursing staff 61-2, 73, 86

obesity 30
occipital lobes 17, 20, 21
occupational therapist 63-5, 66, 93, 101, 119
    home care, 110, 112
oedema, *see* cerebral oedema
oestrogens 27, 46
'old man's friend' 87
optic disc swelling 51
osmosis 76
outcome, after stroke, *see* prognosis
outpatient care 63, 109, 111
oxygen utilization, cerebral 15

papilloedema 51
paralysis 6, 7, 22, 85, 129, 130
    (*see also* hemiplegia)
parietal lobes 17, 19, 39
pencil injury 32
perception 17, 19
Persantin 81, 82
personality changes 14, 16
phlebography 89
physical assessment 94-6
physical examination 71-2
physical tolerance 62, 63
physiotherapy 62-3, 65, 85, 98, 119
    chest 62, 85, 87-8
pineal gland 54
platelet 77, 80-2
polycythaemia 47
pons 124, 129
positron emission tomography 57
Praxilene 78, 87

pressure sores 61, 85
'presumed stroke' 5, 71, 76, 108, 133
prevention, of stroke 26, 31, 103-4, 132-3
prognosis, of stroke 8, 35, 73, 114, 124-31
prostacyclin, PGI_2 81, 82
psychological assessment 17, 92-4
psychological recovery 114-15
pulmonary embolus 87, 89, 90

race, stroke incidence 28
reading 17, 19, 93, 128
'rebound' 77
receptive dysphasia 18
recovery
    long-term, prediction 130-1
    pattern 114
    rate 98, 99, 117
rehabilitation 60, 61, 64-5, 91-7, 99-102, 119-23
    assessment 121-2
    units 98, 100-2, 115, 134

remedial therapy 119-23
research and future 132-4
reversible ischaemic neurological deficit (RIND) 8
risk factors 25-32, 82-3, 132
routine tests 52-4

scurvy 47-8
season of year 29
sensory area 17, 19
sensory inattention 19, 92, 95
sensory system, assessment 95
septum pellucidum 51-2
sex 27, 82, 128
sheepskins 85
sickle cell disease 31, 32, 47
site, of stroke 13, 16-22, 71-2, 93, 124
smell, sense of 16
social assessment 97
social network 97, 99, 110-12
social problems 67, 68, 70, 97, 114-15
Social Services Department 110, 115
social worker 66-7, 111, 112

Index

socio-economic status 28-9
sodium nitroprusside 84
specialists 59-67, 86, 90, 98, 119
spectrophotometry 51
speech assessment 91-2
speech centre 13, 16-18, 19
speech disabilities 18, 65-6, 91-2, 114
   recovery, prognosis 127
speech therapist 65-6, 91, 119
stenosis 23, 37, 56, 58
steroids 75-6
stroke 1, 5
   description 9
   types 33-49, 117
'stroke-in-evolution' 8, 40, 79
stroke unit 59, 133
subclavian steal 43
sulphinpyrazone 81, 82
supportive therapy 84-6
surgery 103-6, 134
Sylvian fissure 104
syphilis 48

temporal lobe 17, 18, 19
therapists, *see specific therapists*
therapy, remedial 119-23, 134 (*see also* rehabilitation)
thrombosis 79, 82, 125
   arterial 81
   cerebral 7, 46
   coronary 36, 70
   deep vein 80, 85, 88-9, 90
thrombus 79, 81, 89, 126
tomography 56, 57 (*see also* computerized tomography)
transient ischaemic attack (TIA) 5, 6, 8, 23, 30, 41-2, 132
   drug treatment 79, 80, 82, 132
   investigations 103

platelets, role in 81
treatment, of stroke
   acute 133
   assessment 116-23
   drug, *see* drug treatment
   surgery 103-6
   trials 133
      drug 116-19
      therapy 120-1, 134
tumours 48-9, 51, 54, 55
   treatment, drug 75, 76

ultrasound 51, 58
unconsciousness 33, 41
'upside-down man' 18, 19

vascular malformation 43-4, 56
'vascular origin' 5
vasodilator drugs 78
verbal memory 17, 19
vertebra, cervical 11, 42-3
vertebral artery 11, 42
vertebrobasilar ischaemia 8, 42-3
vision, loss of 35, 42
visual area 17, 20-1
visual inattention 19, 39, 92, 96
visual system, assessment 96
vitamin C 47-8

walking 62, 64, 98, 114
Warfarin 79, 80
Wernicke's speech area 18, 19
Willis, circle of 10, 11, 125
workshops 64, 101
writing 16, 17, 19, 93

xenon-133 15
X-ray 12, 56, 58
   chest 54, 56, 72, 87, 90
   skull 54